Urban Redevelopment

Urban redevelopment plays a major part in the growth strategy of the modern city, and the goal of this book is to examine the various aspects of redevelopment, its principles and practices in the North American context.

Urban Redevelopment: A North American Reader seeks to shed light on the practice by looking at both its failures and successes, ideas that seemed to work in specific circumstances but not in others.

The book aims to provide guidance to academics, practitioners and professionals on how, when, where and why specific approaches worked and when they didn't. While one has to deal with each case specifically, it is the interactions that are key. The contributors offer insight into how urban design affects behavior, how finance drives architectural choices, how social equity interacts with economic development, how demographical diversity drives cities' growth, how politics determine land use decisions, how management deals with market choices, and how there are multiple influences and impacts of every decision.

The book moves from the history of urban redevelopment, The City Beautiful movement, grand concourses and plazas, through urban renewal, superblocks and downtown pedestrian malls to today's place-making: transit-oriented design, street quieting, new urbanism, publicly accessible, softer, waterfront design, funky small urban spaces and public-private megaprojects. This history also moves from grand masters such as Baron Haussmann and Robert Moses through community participation, to stakeholder involvement to creative local leadership. The increased importance of sustainability, high-energy performance, resilience and both pre- and post-catastrophe planning are also discussed in detail.

Cities are acts of man, not nature; every street and building represents decisions made by people. Many of today's best recognized urban theorists look for great forces; economic trends, technological shifts, political movements and try to analyze how they impact cities. One does not have to be a subscriber to the "great man" theory of history to see that in urban redevelopment, successful project champions use or sometimes overcome overall trends, using the tools and resources available to rebuild their community. This book is about how these projects are brought together, each somewhat differently, by the people who make them happen.

Barry Hersh is a Clinical Associate Professor of Real Estate, teaching graduate courses in property development and coordinating the development program for the New York University Schack Institute of Real Estate, in the School of Continuing and Professional Studies.

Urban Redevelopment
A North American Reader

Edited by Barry Hersh

LONDON AND NEW YORK

First published 2018
by Routledge
2 Park Square, Milton Park, Abingdon, Oxon OX14 4RN

and by Routledge
711 Third Avenue, New York, NY 10017

Routledge is an imprint of the Taylor & Francis Group, an informa business

© 2018 selection and editorial matter, Barry Hersh; individual chapters, the contributors

The right of Barry Hersh to be identified as the author of the editorial matter, and of the authors for their individual chapters, has been asserted in accordance with sections 77 and 78 of the Copyright, Designs and Patents Act 1988.

All rights reserved. No part of this book may be reprinted or reproduced or utilized in any form or by any electronic, mechanical, or other means, now known or hereafter invented, including photocopying and recording, or in any information storage or retrieval system, without permission in writing from the publishers.

Trademark notice: Product or corporate names may be trademarks or registered trademarks, and are used only for identification and explanation without intent to infringe.

British Library Cataloguing in Publication Data
A catalogue record for this book is available from the British Library

Library of Congress Cataloging in Publication Data
Names: Hersh, Barry.Title: Urban redevelopment : a North American reader / edited by Barry Hersh.
Description: New York : Routledge, 2018. | Includes bibliographical references and index.
Identifiers: LCCN 2017010124| ISBN 9781138786400 (hardback : alk. paper) | ISBN 9781138786417 (pbk. : alk. paper) | ISBN 9781315767314 (ebook)
Subjects: LCSH: Urban renewal–North America.
Classification: LCC HT178.N69 U73 2018 | DDC 307.1/416097–dc23
LC record available at https://lccn.loc.gov/2017010124

ISBN: 978-1-138-78640-0 (hbk)
ISBN: 978-1-138-78641-7 (pbk)
ISBN: 978-1-315-76731-4 (ebk)

Typeset in Times New Roman
by Wearset Ltd, Boldon, Tyne and Wear
Cover image T. Lawrence Wheatman

To my wife Jeanne, our daughters Alayne and Michelle, and the memories of my parents Ruth and Phil and my sister Karen

Contents

Urban redevelopment is a growing and challenging field of city planning, design and real estate. In North America, the recycling of underutilized land within communities is both extraordinarily complex and significant. This collection of articles and case studies examines the key aspects of urban redevelopment and how each contributes to modern cities.

Notes on contributors	xii
Foreword	xv
CHARLIE BARTSCH	

Why government plays a vital role in addressing systematic community economic development challenges xv
Lessons from EPA's community and economic development experiences: what can inform a broader inclusive approach? xvi
Identifying and maximizing potential drivers of community growth: what framework, ideas and strategies advance public sector efforts? xviii

1 History and trends 1
BARRY HERSH

History of urban redevelopment and renewal 1
Baltimore as a model 2
Case study: Eastwick, Philadelphia, Pennsylvania 5
Measuring urban redevelopment trends 2017 6
Urbanophile case study by Rod Stevens 14

From urban renewal and slum clearance, urban redevelopment has transitioned to contextual design and neighborhood preservation.

2 Historic preservation 21
BARRY HERSH

The historic preservation of landmark structures, and especially districts, has become a controversial but critical element of urban redevelopment. What is preservation? Crucial for a community versus the rights of property owners to develop larger, more modern buildings is a key debate in many cities.

Examples of adaptive reuse in Toronto: Evergreen Brickworks 24
Toy Factory Lofts 24
North Toronto Station 24
Bethlehem Steel 28

3 Urban design and city form in redevelopment 31
WILLIAM SCHACHT

Introduction 31
Urban design process 35
Parameters 36
Technology and tools for urban design 39
The first mandate: safe, secure and resilient 40
The urban design plan 42

Urban design of redevelopments can be, at best, examples of beautification and creativity. Design can help mold the social and psychological as well as physical and real estate impact of redevelopment. The use of density, land uses, height, waterfront, public spaces and skyline all interact.

Urban design form 46
Design elements 51
Case study: Kohn Pedersen Fox – contemporary global urban design project 55
Case study: design for community crime prevention – defensible space revisited 55
Case study: Rocket Street, Little Rock, Arkansas 57
Case study: Vancouver, British Columbia 58

4 Transportation 62
G.B. ARRINGTON

Urban redevelopment is often transit oriented, exemplifying the generational move away from the auto-dependent suburban lifestyle. Projects frequently emphasize use of not only rail, but also of bicycles and walking. Redevelopment can sometimes utilize, but often upgrades, existing sewer, water and other infrastructure and may offer new services ranging from big data analytics to local internet.

People moving to city shaping 62
Modern streetcars 63
Two different paths to a twenty-first century metamorphosis 64
BART's journey into the twenty-first century 65
Tysons Corner: from Edge City to twenty-first-century city 66
Conclusion 67
Transportation case study: the Pearl District – Portland's largest TOD 68
Development oriented transit 68

Public and private initiatives shaping the Pearl District 69
Other urban infrastructure and sustainability 69
Case study: Denver TOD – the next big thing? 71
Incrementally, then boldly building a regional rail system 71
TOD evolution: from city with transit to transit city 72
Early TOD planning in the Denver Region 73
New tools, new partners and new goal posts 75
Central city riches, suburban focus 76
Prospects for the future 78

5 Parks, open space, arts and culture — 81
BARRY HERSH

Urban redevelopment often thrives near public open space. Rediscovering urban parks, improving access to waterfronts or creating new amenities are often key elements of urban redevelopment. Arts and artists are often early harbingers of revitalization and can play a key role in long-term redevelopment.

Arts and culture 82
Institutions 85
Mini-case studies 85
Gas Works Park, Seattle 85
Dry Gulf Stream restoration at Lamar Station Crossing, Lakewood, Colorado 86
Greenway, Ranson, West Virginia 86
Myriad Botanical Garden, Oklahoma City, Oklahoma 86
Durham Performing Arts Center, Durham, North Carolina 86
Spruce Street Harbor Park 87
Discovery Green, Houston, Texas 88

6 Environmental issues – brownfields — 89
BARRY HERSH

The intensity and infrastructure of cities make them inherently more energy efficient and less polluting than leafy suburbs. Compact and well located cities can also be made more resilient. Remediation of contamination is often an important and beneficial requirement of redevelopment; all of which makes urban redevelopment the smartest form of growth.

Other environmental concerns, noise and air quality 95
Waterfront redevelopment 96
A Leadership and building a team 97
B Approval strategies 98
C Innovative financing 100
D Strategies: site acquisition 102
E Synergy between remediation and redevelopment 103

F Maximizing the benefits of waterfronts and creating true mixed-use for
　　　waterfronts 105
　Case study: Harbor Point, Stamford, Conneticut 106
　Two case studies, Toledo, Ohio 108
　Case study: gas stations 110

7 Revitalizing neighborhoods, housing and social equity   113
　GENEVIEVE LEE CABANELLA

Urban redevelopment inevitably changes a neighborhood, differences in housing type and quality, economic opportunities, amenities and crime prevention often also result in gentrification. How are affordable housing, inclusionary zoning, design standards and other tools used to support residents but still encourage redevelopment?

　History of urban renewal and public housing 114
　Financing affordable housing 116
　Regulations and incentives in urban development 116
　Mixed-use affordable housing 117
　Land trusts, urban agriculture and redevelopment 118
　Innovation in urban revitalization, gentrification 119
　Community engagement 120
　Case study: Camden, New Jersey 121

8 Real estate and capital markets   125
　RICK MANDELL

How real estate development, especially urban redevelopment, has changed and become more challenging as the economy has emerged from the 2007–2010 financial crisis and recessions. While investor goals remain constant, techniques, measures and perceptions change dramatically.

　Funding the gap 128
　Real estate marketing 129
　Developers' perspective 130
　Economic development 130
　Business improvement districts 133
　Redevelopment real estate taxes and liens 133

9 Megaprojects   136
BARRY HERSH

Megaprojects, often urban redevelopments, are an increasing share of development. Most are major public-private partnerships, involving government approval and support, often of infrastructure and cleanup. Megaprojects often include major facilities such as stadiums, parks and transit hubs. Often these projects include stadiums, arenas, convention centers and other major public features.

Case study: Atlantic Station, Atlanta, Georgia 137
Case study: Manhattan West Side, the High Line and Hudson Yards 138
Rebuilding Detroit 142

10 The urban redevelopment process: putting it all together   149
BARRY HERSH

Key factors in success or failure 154

What can be learned by failure of projects and from declining cities? Defining successful urban redevelopment, identifying successful and innovative strategies for communities, and the role of urban redevelopment in creating sustainable cities.

Bibliography   159
Index   162

Contributors

Editor

Barry Hersh, AICP, has extensive experience in urban development as a practitioner as well as academic. He is a Clinical Associate Professor at the New York University Schack Institute of Real Estate and Chair of the MS in Real Estate Development Program. He has been actively engaged in urban redeveloped as a planner, real estate developer, community board member and observer, in North American cities from Stamford, CT, to Toledo, OH. He has written extensively on various aspects of urban redevelopment, including a national study of Waterfront Brownfield Redevelopment published by the NAIOP Foundation who also named him a 2014–2017 Distinguished Fellow. His writings on redevelopment include "Brownfields on Three Continents" published by the American and Asian Real Estate Societies and presented in Shanghai and numerous articles in *Brownfield Renewal* and other professional publications. His New York City Brownfields Study, recognized in 2006 for Outstanding Scholarly Achievement, was utilized in PlaNYC 2020. He holds a Master's Degree in Urban Planning from New York University's Wagner School and has a Bachelor's in Urban Studies from CUNY. He is a Certified Environmental Planner and member of the American Institute of Certified Planners. He is also a board member of New Partners for Community Revitalization, the Redevelopment Institute, and since 2014, he has been a member of the City of Stamford Historic Preservation Commission.

Contributors

G.B. Arrington is the principal of GB Placemaking based in Portland, OR. He is one of the world's most respected innovators in Transit Oriented Development (TOD). His focus is in strategically forging the link between transit and development to catalyze community revitalization, resiliency and place-making. He brings the insights and knowledge garnered from four decades of successfully shaping public policies, land use plans and transit projects at the scale of the region, the corridor and individual sites. He was previously with Parsons-Brinkerhoff and TriMet and earned his degree from Herriot-Watt University in the United Kingdom.

Charlie Bartsch was Senior Advisor for Economic Development to EPA Assistant Administrator Mathy Stanislaus, charged with promoting interagency and public-private financing partnerships to spur land revitalization and site reuse. Among his other duties at EPA, he works closely with the EPA-DOT-HUD Partnership for Sustainable

Communities, advises the Office of Brownfields and Land Revitalization on area-wide planning and auto communities revitalization financing strategies, and is taking a leading role in developing the agency's EJ2014 strategy, addressing equitable development concerns in environmental justice communities. He has also been an EPA point person on the White House "Strong Cities/Strong Communities" recovery initiative and on the Joint Initiative on Urban Sustainability Initiative with Brazil. Prior to his appointment at EPA, he was Senior Fellow at ICF International, where he served as ICF's brownfields and smart growth policy expert. Before that, he was Director of Brownfield Studies at the Northeast-Midwest Institute in Washington, DC, a public policy center affiliated with the bipartisan Northeast-Midwest Congressional and Senate Coalitions. Over the past 20 years, his focus has been on brownfield and community redevelopment/reuse strategies and financing, and he is recognized as one of the nation's leading authorities on these issues. He has provided training and technical assistance support in more than 200 communities in over forty states. He has written numerous reports and other publications on these issues, including the pioneering Coming Clean for Economic Development; New Life for Old Buildings; Coping with Contamination: Industrial Site Reuse and Urban Redevelopment; and two annual reference resources, Brownfields "State of the States" and the Guide to Federal Brownfield Programs. He most recently co-authored "Financing Strategies for Brownfield Cleanup and Redevelopment, Recycling America's Gas Stations, and Financing Renewable Energy Projects on Contaminated Properties – Strategies and Options." He often testifies before Congress on issues of economic development, most recently on HUD brownfield financing innovations and brownfield tax incentives. Prior to his service at EPA, he was chair of the National Brownfield Association's Advisory Board, chair of GroundworksUSA and on the editorial board for the Bureau of National Affairs. In 2001, he received the International Economic Development Council's Chairman's Award for Outstanding Service for ten years of work on brownfield policies and legislation. He received his Masters in Urban Policy and Planning from the University of Illinois-Chicago and his B.A. in political science from North Central College in Naperville, Illinois.

Genevieve Lee Cabanella currently works as a Community Investment Analyst at the Federal Home Loan Bank of New York in the Affordable Housing Program. Her role at the FHLB-NY AHP involves analyzing the financial and cost feasibility of various multifamily affordable housing developments, including mixed-use, mixed-income, supportive housing, senior housing, homeless shelters and even artist housing. She previously worked for non-profit community development organizations and supervised pre-development, construction and completion of affordable housing at New Community Corporation and HANAC, Inc. She obtained her Master's degree at the New York University Schack Institute of Real Estate and Bachelor's degree at Boston College Carroll School of Management.

Rick Mandell, Esq., has experience as a real estate developer, attorney and investment banker. Based in Colorado and originally in Florida, he has been involved in all phases of real estate development and financing, currently providing financial intermediary consulting services for private equity funds and banks in the land, homebuilding and deep value real estate sectors across the country providing debt and equity placement services and homebuilder finance. He is a frequent speaker on real estate capital markets for the National Association of Home Builders and comparable professional organizations. He holds his law degree from University of Florida – Frederic G. Levin College of Law.

William Schacht, AIA, RA, is currently active in private consultancy including involvement with New York New Visions, a coalition developed to provide guidelines, standards and recommendations for New York City's Ground Zero. He has over forty years of experience in the practice of architecture and urban design, with a current focus on educational and professional consultancy services regarding design, projects management, construction administration and firm development. Personally responsible for an array of award-winning projects in commercial, institutional, research and development, residential and industrial market sectors, he holds a Bachelor of Architecture from the University of Illinois, a Master's of Science in Urban Design from Columbia University and a Master's of Urban Design from City University of New York. He is a registered architect in the states of New York and Michigan and has been awarded over seventeen international, national, state and New York City design awards for his work, and he is an architect and urban designer who has worked with some of the major New York-based design firms. His recent work includes urban redevelopment in his hometown in Queens, New York.

Rod Stevens is a noted urban commentator and a seasoned real estate professional specializing in the positioning and planning of innovation districts, tech centers and other sophisticated places for research, development and production. He is a trusted advisor to institutions and investors on the strategic use of their real estate assets to develop new business and attract talent and investment. He has more than 30 years of experience leading teams of skilled professionals in identifying new and emerging needs, programming and positioning projects, and negotiating the approvals, financing, and use agreements. He produces new, successful and creative development for the businesses that use them, the investors that own them and the communities that surround them. Mr. Stevens holds a B.A. from Stanford University, was a Tuck Scholar and received his MBA from the Tuck School of Business at Dartmouth College. He currently lives on Bainbridge Island, Washington.

Foreword

Charlie Bartsch

Government has long concerned itself with local and regional community economic development, especially the urban redevelopment discussed in this book. Executive branch officials at both federal and state levels pursue economic development initiatives to spur job growth, increase income for local residents, generate business start-ups and transitions and expand the tax base – all aimed at improving the quality of life, stabilizing communities, enhancing middle class opportunities and reducing poverty. In practice, these efforts have had mixed results; in some cases, government has perpetuated bad local planning through funding that has isolated neighborhoods, channeled capital investment away from distressed areas, nascent small businesses and people that need it, and facilitated development approaches that have excluded the stakeholders with the most to gain – or lose.

We have also seen some important successes, certainly at the federal level, which could serve as practical and effective models for future efforts. From my vantage point as economic development advisor at the US Environmental Protection Agency – focusing on interagency and intergovernmental community redevelopment initiatives – I can affirm that EPA has, through its land-based programs, built a solid track record of proactively promoting community revitalization efforts. EPA's brownfield program, in particular, is based on the notion of improving the physical and social environments for a broad range of community stakeholders that advances an inclusive economic development approach that addresses such key livability principles as a living wage, enhancing access to jobs and commercial/social services, and support of the middle class.

Why government plays a vital role in addressing systematic community economic development challenges

A community's economic development and growth depends upon the independent actions of many players, the overwhelming majority of whom operate in the private sector. A key public sector function, therefore, is establishing the right climate that encourages these private sector actions to take place so that they have the most positive impact on the leveraging of resources to advance broader social, community, and environmental goals – essentially more inclusive economic development.

In many communities – big city downtowns, smaller community main streets, as well as neighborhoods in transition – former economic centers, often industrial legacy sites, can be identified. Often characterized by poverty and blight, these are the communities that need federal attention to recover; these are also the communities that EPA has addressed in recent years, through its land revitalization and brownfield programs, which have played a significant role in advancing economic development and social justice objectives as part of the revitalization process. The types of successes that EPA has seen, the recovery strategies it has promoted and the leveraging and partnerships it has fostered are reflected in the different elements of this book.

The myriad of problems that today face distressed, economically disadvantaged communities is reflected in the content of this book and has been well documented elsewhere. Urban sprawl and central city abandonment were fueled by past siloed federal spending, which subsidized industrial and commercial economic development and housing construction away from urban and rural centers. Private investment followed the public sector path, often led by the federal government. At the national level, few programs focused on rehabilitation and restoration; the focus was on new development in previously undeveloped areas. Many state programs took similar approaches. This resulted in stratification of population based on race and class. This situation has further been exacerbated by politics, which also has stratified along urban, suburban and rural lines. Prior generation bipartisan and regionally diverse support for social investment, such as in affordable housing and infrastructure, has been impacted by this stratification.

These systematic problems cannot be addressed by government that continues to address social and economic issues in silos, that does not evolve its efforts in ways that reflect local needs that communities themselves have defined through a participatory planning and problem-solving process and link them to preferential access to economic development resources. Local leaders around the country have recognized that broader revitalization of their communities is linked to the reuse of assets that have been used before, especially sites, facilities and properties, and infrastructure.

Even with their many problems, virtually every community has advantages that can be nurtured and built upon to achieve economic benefits for all local residents. Main streets, downtowns and individual neighborhoods have the advantage of built infrastructure, transportation access and local work forces that are critical to advance economic development. What they need are a set of policies that reverses the fractured public economic development/infrastructure spending that has resulted in displacement and sprawl – specifically, programs that tie project implementation funding to good planning, make it easier for communities to gain access to the range of state and federal assistance that they may need to more systematically pursue their revitalization visions and provide real opportunities for everyone in the workplace and marketplace. Recognizing this is the challenge – and acting on it represents a great opportunity for governments at all levels.

Lessons from EPA's community and economic development experiences: what can inform a broader inclusive approach?

First conceptualized in the Clinton Administration, EPA's brownfield land revitalization initiatives have more than 20 years of experience characterized by creative community revitalization and productive private sector involvement. A pioneering environmental effort aimed at integrating private sector investment needs with responsible environmental stewardship, brownfields – defined as sites and properties where real or perceived contamination inhibits the productive reuse and revitalization of the property – have evolved into a critical community economic redevelopment strategy. Brownfield sites are critical because their locations in the heart of communities makes them anchors to catalyze broader community revitalization. It provides common ground for a range of public, private, non-profit and neighborhood interests working towards community betterment – more and better jobs, increased business and service opportunities and a better quality of life.

EPA's brownfields program has evolved in tandem with the changing nature of public-private partnerships aimed at community development in economically and socially distressed communities, through multi-stakeholder inclusiveness – a critical building block of a community revitalization strategy. The brownfield program assists communities in responding to local economic revitalization challenges by supporting preparation of planning and project implementation strategies that leverage new business and community investments that lead to jobs and other benefits for all local residents.

The brownfield program's economic development track record is impressive: since the program's inception, it has been credited with creating nearly 106,000 jobs, stimulating numerous new business opportunities, increasing residential property values as much as 12 percent, while reclaiming and bringing new life to more than 49,000 contaminated acres. It has led to project partnerships involving more than two dozen different federal agencies, and more than 100 state agencies. Brownfields is also one of the most cost-effective federal investments – each EPA program dollar has leveraged nearly $18 in other investment.

Given its on-the-ground project implementation focus, EPA's brownfields program can show policy makers how to look practically and broadly at "inclusive economic development," as a strategy to connect private capital with redevelopment needs and opportunities in distressed areas, in order to promote residents' income growth and diversification. Its partnerships, leveraging and cross-sectoral impacts serve as a solid model on which to base public sector community and economic development policies broadly. And they do so in a way that invites private sector support of key environmental objectives.

Simply put, although brownfields initially focuses on the real or perceived environmental contamination on formerly used properties, the program has had a sizable impact on community development outcomes because the location of brownfield properties is typically in downtowns and economic centers, which have been anchors to catalyze community revitalization. To date, EPA's brownfield program resources have leveraged more than $22.6 billion in other investment; many communities now see preparing brownfields for reuse as the vital first step towards achieving economic development recovery and growth. The context for the brownfield concept, and the lessons EPA has learned from carrying it out on the ground, offers a good example of inclusive community development that addresses poverty, and brings job creation, economic development and environmental benefits. Brownfield sites are the rare circumstances where government can invest in local land use decisions and development, leveraging such investments for a number of broader social and economic goals.

Identifying and maximizing potential drivers of community growth: what framework, ideas and strategies advance public sector efforts?

Going forward, both state and federal governments should consider a range of approaches to strengthen local economies in ways that promote recovery of distressed or abandoned areas in cities and towns of all sizes as part of a national economic development goal, and bring benefits for a full range of community residents and stakeholders. Many of these ideas are reflected in chapters throughout this book. For example, several of the authors note that, given the dynamics of community revitalization, policy makers need to realize that the communities themselves will be at different stages of readiness to define an economic recovery and growth vision, and then access and deploy implementation resources to carry out that strategy. Recognizing these variations is a key factor, and integrating them into an economic policy framework plays an important part in the efficient and effective shaping of polices aimed at helping communities and promoting an inclusive redevelopment strategy; this has been a key lesson of EPA's site revitalization efforts.

There is no doubt that government decisions, policies, investments and activity significantly influence the economic environment of an area when it comes to business climate, site use, appropriate training and education, infrastructure investment and upkeep, taxes and regulations, and the provision of public services. And, depending on how ready a community is to advance development efforts, government may also take more targeted actions for community development to assist vulnerable populations, address environmental concerns, support small businesses and help workers. Several considerations, reflected in the chapters of this book, will influence how public sector involvement should be defined and how key policy assumptions should be applied. These include addressing information shortcomings by providing usable and appropriate information that may be lacking in critical situations – for investors searching for opportunity, for businesses looking for suitable trained workers, for communities seeking to redeploy vacant and abandoned sites for new uses, for entrepreneurs needing start-up capital or business know-how, for businesses using outmoded production or management techniques, or for workers hoping to hone their skills and knowledge for emerging occupations and industries.

In short, redevelopment strategies succeed when communities see accomplishments. The ideas and suggestions in this book well define the next generation of public sector strategies and approaches for local economic development that includes all stakeholders and stems from constructive public-private partnerships. As we have learned from our considerable successes with EPA's brownfields program, government has both a strong interest and legitimate role to play in community economic development efforts aimed at strengthening, revitalizing and growing cities and towns, and improving the overall quality of life for local residents. Government helps spur community development through investments in public infrastructure, the provision of public goods and services, and targeted assistance to industries, businesses and workers – education, social services, R&D and many other services. Public – especially federal – development efforts work best when initiatives build upon community potential and strengths to improve the long-run outlook for growth. The best community and economic development approaches will flow from careful and realistic analysis of a local or regional economy and its potential for growth.

Acknowledgments

I would like to thank the many who helped and supported me in the writing and editing of this book, starting with the volunteer contributors who graciously gave their time and expertise. There were several New York University SPS Schack Institute of Real Estate graduate students: Claire Han, Omar Hamani and Shawn Dacey, who contributed to the research effort. My colleague, T. Lawrence Wheatman, generously provided several valuable photographs, including the cover image. Many other friends and colleagues at New York University, in the redevelopment and brownfields community, and beyond, provided ideas and discussion. Finally, thank you to my friends and family who assisted me throughout the long and arduous process.

1 History and trends

Barry Hersh

History of urban redevelopment and renewal

Cities have always reinvented themselves; urban land has constantly been redeveloped. Today, urban redevelopment is an ever-growing phenomenon in North America, more and more urban communities are reinventing themselves; land that was built upon, utilized, then as times change, structures are demolished and new ones built, land was reused over and over. There is a worldwide heritage of city rebuilding; visit Jerusalem, and see layers of history, as each conqueror built their own house of worship atop the old. Walled cities expanded, the defensive value of walls diminished, but the castle and the core of the city remained.

It was Baron Haussmann (1809–1891), on behalf of Emperor Napoleon III, who gave new meaning to government led urban redevelopment, by building wide boulevards and public places, demolishing existing structures in their path and making Paris a modern city.

In North America, after World War II, in part due to government highway and mortgage policies, suburban growth exploded while the center of many cities struggled with economic losses and racial discrimination. By 1948, the US federal government created a

Figure 1.1 Public housing on Urban Renewal site
Source: Pruitt-Igoe St. Louis Housing Authority.

further extension of government city redevelopment called Urban Renewal, two perfectly good words that are still stigmatized by the "federal bulldozer" approach of that program, clearing acres of urban neighborhoods designated as blighted and taken by eminent domain, to build "superblocks" often of public housing and sometimes with new private development. Robert Moses the long-time "Power Broker"[1] of New York came to epitomize this approach, building highways, public housing and parks often by demolishing large swathes of often low-income and minority communities.[2] In the 1960s, Jane Jacobs, the author of *The Death and Life of Great American Cities*,[3] became Moses' nemesis by stopping a proposed highway through Greenwich Village and giving voice to the "ballet of the sidewalk" and the virtues of historic neighborhoods.

By the late 1960s, urban renewal was unpopular and, in 1974, was superseded by the Community Development Block Grant Program. The new construction of low-income family public housing was partially replaced by the Section 8 voucher program and also by tax credit programs for historic preservation and low-income housing.

Urban renewal efforts in the latter half of the twentieth century had successes and failures, developments that did revitalize some cities, but others failed. While there were many great individual city projects, ranging from Boston's Faneuil Hall to San Francisco's Buena Vista, cities often struggled, even a decade after the turn of the century; suburban growth, especially in the sunbelt, continued to dominate. Federal efforts were curtailed, and much of government urban regeneration was left to states and cities themselves. The 2004 *Kelo v. New London* United States Supreme Court[4] actually upheld, by a 5–4 vote, the use of eminent domain to promote economic development; the political storm that ensued marked the major shift away from government use of eminent domain towards more contextual, public-private partnership forms of community redevelopment. In the 2017 Supreme Court case, Murr v Wisconsin, Justice Kennedy's decision proposed a complex analysis of the parcel, its physical characteristics and economics in determining regulatory taking.

Despite all the efforts to promote so-called smart growth, it was not until after the 2007–2010 financial crisis and ensuing "Great Recession" that a new pattern in urban redevelopment became clear. While federal efforts were more limited, local and state governments aggressively pursued economic development for both job and tax revenue growth. Economic drivers had changed dramatically: it was technology, media, education, arts and health care that employed the new millennials, and it was an amenity-rich, transit oriented and more urban lifestyle that now competed for the next new thing in technology and quantitative finance. Perhaps most importantly, communities are leading their own reinvention, envisioning changes, using land use and environmental planning, and working with private developers, as well as government, to effect change. While changes in administration matter, the pattern, led as much by communities, local and state government as federal programs, continues beyond election cycles.

Moving from history to urban redevelopment in the twenty-first century first is a look at trends in urban redevelopment as of 2015, especially the varying views as to the extent of a new urban renaissance as compared to more suburban preferences. This is followed by a "case study" an idiosyncratic perspective by Rod Stevens that includes a notable time chart of urban redevelopment efforts.

Baltimore as a model

It is hard to talk about American urban redevelopment without discussing Baltimore. The redevelopment of the Inner Harbor became a planning model for downtown and waterfront

Figure 1.2 Transit Oriented Development has become one of today's approaches to urban redevelopment
Source: National Transit Institute.

redevelopment. Starting in 1958 with the Charles Center downtown project, this urban renewal project then transformed the adjacent decrepit waterfront of deteriorating piers into a major tourist attraction. On July 4, 1976, eight tall sailing ships from other nations visited Baltimore's Inner Harbor, attracting a huge number of tourists. This interest helped spur the development of other attractions – including the National Aquarium and Maryland Science Center. The Rouse Company's Harborplace festival marketplace opened on July 4, 1980. The nearby Baltimore Convention Center and Hyatt Regency Baltimore Hotel added to the services and resulted in population density and visitors. In the years that followed, Baltimore worked to expand its redevelopment efforts, focusing on near downtown neighborhoods such as Federal Hill and Fells Point, though there were efforts elsewhere in the city. Oriole Park at Camden Yards, adjacent to downtown and opened in 1992, became another model of creative historic adaptation and the use of sports and entertainment venues as catalysts for redevelopment.[5] In many ways, Baltimore became the model of successful urban renewal.

Baltimore's success was largely credited to exceptional leadership, starting with Jim Rouse, successful commercial mortgage and shopping center entrepreneur, who became the private sector champion of the development effort. After retirement, Rouse went on to create Enterprise Communities, one of the largest non-profit developers and financiers of affordable housing in the country. Donald Schaefer, Mayor of Baltimore from 1971–1987, Governor of Maryland from 1987–1995 and Maryland Comptroller from 1999–2007, became legendary. Many of the designers, David Wallace of WRT, James and Jane Thompson and Martin Millspaugh, the redevelopment director, went on to consult and

Figure 1.3 Baltimore Inner Harbor
Source: Baltimore.org.

design. Baltimore became a model of urban revitalization exported and emulated in Boston, New York and other cities around the US and the world.

Yet the Baltimore of David Simon's HBO series *The Wire*, with drug dealing, political payoffs and crime, still exists. Some neighborhoods including Sandtown-Winchester and East Baltimore remained stubbornly mired in decline despite redevelopment efforts. These neighborhood issues became more of the focus. In 2012, Ronald Daniels the President of Johns Hopkins said "If EBDI (the East Baltimore Development Initiative) fails, then my presidency at Hopkins fails."[6] Hopkins effort to improve the increasingly dangerous adjacent East Baltimore neighborhood is substantial, totaling over $1.2 billion. Started in 2003, EBDI takes pride in having community residents on its board and having properly relocated only 534 families in preparing its 31 acre site. Projects include a new Hopkins Biotech Center, being developed in conjunction with Forest City (large Cleveland-based firm whose projects include Pacific Park/Barclay Center in Brooklyn and former Stapleton Airport in Denver). Another arm of Johns Hopkins moved into an 1876 renovated police station. There have been objections, including a group led by a former Hopkins doctor, but the project is proceeding somewhat slowly with new Hopkins facilities, graduate student housing, and both new and affordable housing for residents completed or under construction.

In 2015 Baltimore, or more precisely some neighborhoods in the city, exploded with the death of Freddie Gray while in police custody. Despite all the redevelopment, a series of generally well-regarded mayors white and black, and a police force that is over half minority[7] – poor, predominately black communities such as Gray's Sandtown-Winchester felt isolated and not benefiting from all these efforts. Baltimore is still recovering, but there are $2 billion of new projects, mostly near downtown and Hopkins, moving forward.

Another noteworthy new effort is in the Port Covington neighborhood, led by local success story Kevin Plank founder of Under Armour, to redevelop a neighborhood including a new headquarters. Primarily a privately funded effort, the project may involve Tax Increment Financing (discussed in Chapter 8). As in many other metropolitan areas, some of the issues and opportunities expand into nearby suburbs.[8] Yet Baltimore is still trying to understand how its redevelopment efforts can reach more of its citizens.

There are many lessons to be learned from Baltimore, as a mixture of both success and failure. First, downtowns can be renewed, strong reuse concepts, government and business leadership, financial commitments and great design can make a real difference. Rebuilding the heart of a city is crucial, but so is work in the toughest neighborhoods. New buildings and attracting more jobs will improve an area – but will not reach every neighborhood, much less every person. Educational programs, support for the most disadvantaged and programs aimed at reaching individuals can aid those not helped by shiny new buildings. Noted planner Bill Fulton commented,

> In large part, the answer boils down to the old people-versus-place question in urban policy: Do you focus on improving struggling neighborhoods in the hopes that everyone in the neighborhood will be better off? Or do you focus on helping people get a leg up, even if it means they leave the neighborhood?[9]

Redeveloping and uplifting a community, especially a downtown, can be thrilling and truly beneficial, but programs aimed at individuals and poor neighborhoods are also needed.

Case study: Eastwick, Philadelphia, Pennsylvania

The sixty-year-history of redevelopment of the Eastwick neighborhood in Philadelphia, Pennsylvania, tells the story of moving from top down urban renewal to eventually increased community engagement and environmental justice. Eastwick is a low-lying modest community in the southwest portion of Philadelphia, bordering I-95 and the Philadelphia International Airport.

In the early 1950s, President Dwight Eisenhower asked David Reynolds, the head of Reynolds Metals Company, which Eisenhower had helped grow to meet wartime aluminum demands, to become involved in urban renewal. Reynolds, operating through various corporate entities became involved in a number of cities, including Syracuse, Providence, Kansas City and Philadelphia. In 1954, Reynolds formed the New Eastwick Corporation with local builder Berger, designated urban renewal developer, and a plan was prepared by noted architect Constantine Doxiadis, working with famed Philadelphia planner Edward Bacon. This plan was seen as different from other urban renewal efforts, "penicillin not surgery" and utilized the design principles of Clarence Stein and Henry Wright, exemplified in Radburn, NJ, specifically the separation of vehicular and pedestrian traffic.[10] The project was delayed with needed flood controls, infrastructure and partnership issues, breaking ground in 1961. Eminent domain was used to take over 2,000 private properties and displacing more than 8,000 residents in the racially integrated community, later critiqued by Amy Laura Cahn as an abuse of the "blighted" designation[11] and like nearby Chester, Pennsylvania raising environmental justice concerns. The partnership with Berger did not work well, ending in litigation, and a larger local builder, Korman Companies, was brought in to the project.

Over time, an extensive road network was built and then updated, more than 3,000 units were built along with two retail centers, as well as forty industrial properties primarily by

Korman, and parcels near the Philadelphia airport were spun off for three hotels; the latter was among the more financially successful aspects of the project. In 1972, over 1,000 adjoining marsh acres were designated the National Wildlife Refuge at Tinicum, named after Senator John Heinz III after his death in 1991. The Eastwick project, taken over by Korman, largely stalled, plagued by sinking houses, flooding issues and the impact of two adjoining landfills, Lower Darby Creek and Fulcroft, designated as US EPA Superfund sites.

The Urban Renewal Plan was updated periodically (1982 and 2006), but it was not until the original urban renewal agreement approached expiration after sixty years, and most importantly, community activism increased that a new era began.[12] The Eastwick SEPTA (Southeastern Pennsylvania Transportation Authority) station on the airport line opened in 1997 – a condition set by PNC Bank if it was to keep its operations center in the city – has been a big plus for the neighborhood, which for years had to depend on just the Route 37 bus for public transit. There is also now a new trolley transit loop.

In 2012, Korman requested zoning approval for an additional 772 unit development on the remaining 128 acre open space adjoining the Heinz Wildlife Refuge. This plan was strongly opposed by the Eastwick Friends and Neighbors Coalition, working with attorney Amy Laura Cahn of the Public Interest Law Center of Philadelphia, and an agreement was reached that rights to the remaining 128 acres would instead be turned back to the Philadelphia Redevelopment Agency, which has pledged a community-based planning effort. There is a lot of work still to be done in Eastwick as the transition from top down to community-based redevelopment continues.

Measuring urban redevelopment trends 2017

There is an ongoing debate as to how much North America is moving away from six decades of suburban growth and urban decline. In the past decades, the wind of market forces are at the backs of urban redevelopment. Leigh Gallagher, in her book *The End of the Suburbs*[13] makes statements such as "Millennials hate suburbs," "households are shrinking," "we are eco-obsessed" and "the suburbs were poorly designed to begin with." Noted

Figure 1.4 McMansions under construction, Marlboro Township, MD
Source: Wikiwand.

urban scholar, Chris Leinberger wrote, in 2008, a famous article with a picture of suburban McMansions asking "The Next Slum?"[14]

On the other side of the discussion, Joel Kotkin has noted studies that have shown a continued preference for single-family homes. The most recent, January 2015, by the National Association of Home Builders, ranked choices of "suburban, rural and city center" in that order.[15]

Kotkin goes on talk about "villages" as places that combine the privacy of single-family houses with access to public spaces and amenities.[16] There are many variations on this theme, for example some real estate professionals were surprised when three high-rise residential towers in New Rochelle, Westchester County, New York, were successful in the mid-2000s. There were enough renters who valued being directly next to the train station and thirty-five minutes from Grand Central – but who could have a car, enjoy Westchester's amenities and found enough restaurants and a new retail center in New Rochelle. Around the Harrison, New Jersey train station, just outside of Newark, a new mixed-use TOD is under construction. G.B. Arrington describes the extreme makeover of Tyson's Corners from poster child for suburban sprawl to transit oriented community. In Richmond, California, a "greyfield" (a dysfunctional older suburban shopping center) is being replaced with the expansion of a Kaiser Permanente hospital, the reuse of such greyfields for multi-family housing, retail and mixed-use has been a national trend.[17] The choice is not just between suburbs and center cities; it is between communities that offer transportation options and urban amenities and those that don't, and urban redevelopment today is about recreating places that offer variety of housing choices, transportation alternatives, urban amenities – and old-fashioned monocultures.

Zoning and land use planning have changed also; rather than segregating uses, many zoning ordinances now encourage mixed-use, encouraging the urban vibrancy that is the goal of many redevelopment projects. There are a range of new zoning tools, such as

Table 1.1 Characteristics of urban redevelopment as compared to suburban development

	Urban redevelopment	Suburban
Location	City, inner suburb	Outer ring
Transportation	More transit Short streets Parking maximums, structured	More driving Long, curvilinear streets Parking minimums, surface
Land	Previously utilized Smaller blocks with connections Multiple land uses	Vacant or farmland Landscape buffers Single land use
Environmental	Brownfield	Greenfield
Infrastructure	Existing, may need upgrade, ecological	New
Density	Higher Compact, FAR over 1.0	Lower Spread out, FAR less than 0.5
Housing type	Multiple family	Single family
Retail	Place-making	Shopping center, big box

Source: Modified from Barrington-Leigh, C. and A. Millard-Ball, "A Century of Sprawl in the United States," *Proceedings of the National Academy of Sciences of the United States of America*, June 2015, 112(27): 8244–8249. http://dx.doi.org/10.1073/pnas.1504033112.

form-based zoning and overlay districts that can help guide urban redevelopment. Some municipalities have created special districts for unique redevelopment opportunities.

Actually measuring trends requires multiple sources; survey preferences and buying patterns, choices by geography, building type and community. Creating a dichotomy to try and understand trends is illustrated:[18] Demographic changes provide some of the strongest support for the trend towards urban redevelopment. American family size continues to decline with significant increases in the number of people living alone. For the first time in US history, more than 50 percent of adults live alone. This mitigates against the traditional, family oriented single-family home suburban community and towards urban places with more amenities and adult activities.

The demographic trends can also be seen in the marketplace and demonstrated using several different measures. The chart below illustrates the reversal that has occurred, with city growth surpassing suburban growth for the first time in many decades.

Another very recent study utilizing census data that supports including the June 2015 "A Century of Sprawl" by Christopher Barrington-Leigh and Adam Millard-Ball.[19] This research presents a high-resolution time series of urban sprawl, as measured through street network connectivity, in the United States from 1920 to 2012. By this measure, sprawl started as far back as when cars were first becoming common, and far earlier than the interstate highway systems, and was dominant until the mid-1990s. Over the last two decades, however, new streets have become significantly more connected and grid-like; the peak in street network sprawl in the United States occurred around 1994. By one measure of connectivity, the mean nodal degree of intersections, sprawl fell by approximately 9 percent between 1994 and 2012, as noted in "A Century of Sprawl."[20]

No one has more famously illustrated the workplace changes that have changed lifestyles than Richard Florida. The economist and author of the best-selling *Rise of the Creative Class*, and the more recent *Rise of the Creative Class Revisited*, tracks the hugely significant workplace trends towards more service and fewer industrial jobs, and the growth

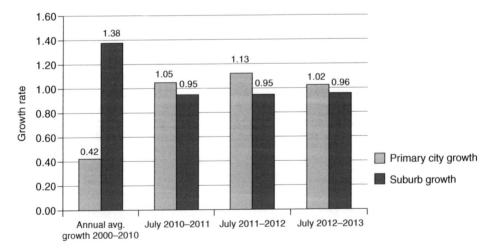

Figure 1.5 Primary city versus suburban growth

Source: Frey, William, *The Atlantic*.

Note: Metropolitan areas over one million population.

History and trends 9

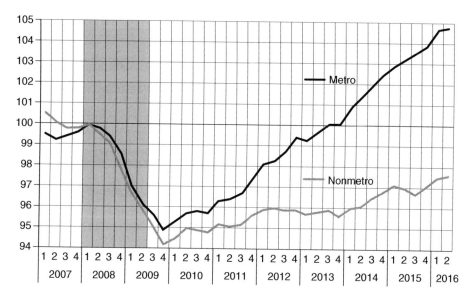

Figure 1.6 US employment, metro and nonmetro 2007–2016 (quarterly)

Source: United States Department of Agriculture Economic Research Service using data from Bureau of Labor Statistics, Local Area Unemployment Statistics (LAUS).

Notes

Employment index (2008 Q1 = 100). Data are seasonally adjusted. Shaded area indicates recession period.

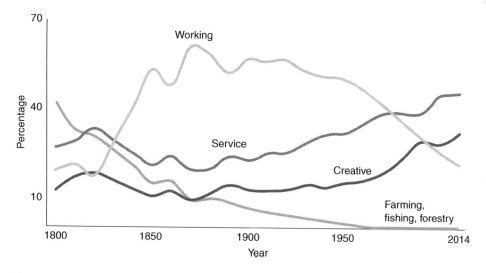

Figure 1.7 Rise of the creative economy

Source: Create Class Group.

of science, technology, education and media. These new types of jobs go along far more with an amenity-rich more urban lifestyle than the traditional suburb.[21] In his 2017 book, the New Urban Crisis Florida notes that economic development does not always bring equity and that every city's urban redevelopment is different.

The millennial generations watched their parents drive increasing commuting times between work and home. The rate of increase in driving trips and time has slowed dramatically, and many young people want to walk, bike ride or take transit to work. As G.B. Arrington describes in Chapter 4, transportation shapes cities and the change in travel choice, based partly on the costs of driving but also on personal choices, which is promoting more urban redevelopment.

Perhaps the ultimate expression of consumer preferences is in the real estate market, where those who rent and buy space make clear choices. As Rick Mandell describes, single-family home builders are facing numerous headwinds, including market changes, as the nation emerges from the Great Recession. As shown by the National Association of Home Builders, apartment construction has out-paced homes in most markets across the United States. Moreover, rental values have grown more sharply than values for single-family homes.

Looking at commercial real estate markets, many suburban office parks, even in the Atlanta region, one of the poster children for suburban sprawl, have seen 20 percent and up vacancy rates and virtually no new construction, while cities and older suburbs centers offering office, transit and residential options packed with amenities, such as long over-looked Hoboken, NJ, are doing better. Not only is the contest between suburban versus Central Business District (CBD) office spaces more competitive in many markets, any amenity-rich, transit-served locations, whether inner suburban, outer part of central city or former "edge city," are competing in the struggling office market. It is the human capital[22] that is critical to today's fast-growing enterprises, and those (mostly young) humans are as likely to opt for

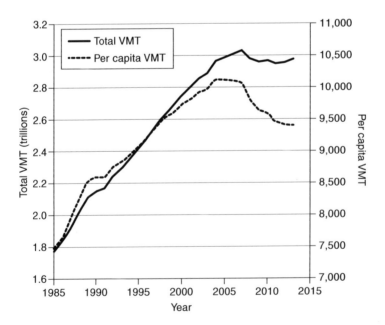

Figure 1.8 Vehicle miles per capita changes

Source: Federal Highway Administration.

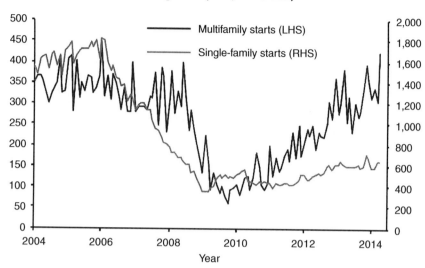

Figure 1.9 Single and multifamily housing starts
Source: National Association of Home Builders.

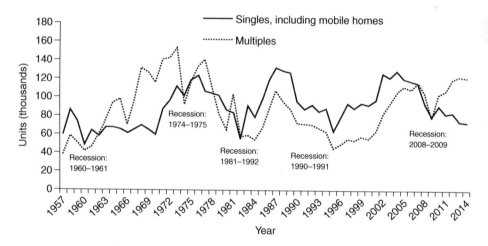

Figure 1.10 Similar pattern of single family as compared to multifamily in Canada
Source: Statistics Canada.

a funky urban community than for a squeaky clean suburb. It is also important to note that today's urban redevelopment occurs in suburbs, mostly older suburbs, so the trend is not completely reflected by suburban versus CBD statisitics. Suburban communities from Cathedral City, California,[23] to Silver Spring, Maryland, and relatively small cities such as Asheville, North Carolina, have successful redevelopment projects.

One of the newer discussions about urban redevelopment in terms of equity is the choice between programs that focus on individuals and those that focus on communities.

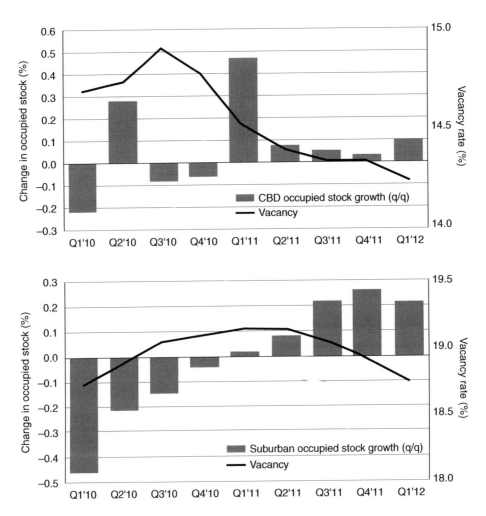

Figure 1.11 Suburban versus CBD occupancy
Source: REIS.

An example of an individually focused program would be recent efforts to provide scholarships for high achieving students from very low-income families. The alternative as noted by Harvard professor Robert J. Sampson and others are programs, including urban redevelopment, that aim at improving an entire community.[24]

Another measure, real estate values, also suggest that more walkable communities have higher commercial property prices, as measured by Real Capital Analytics and WalkScore®.[25] Note that the prices vary in part by CBD as compared to suburban, but more important is walkability. Human resources is the driver and most expensive cost of many of today's growing companies, and these companies like their younger employees are leaning towards more compact, walkable communities. General Electric recently decided to move from a Fairfield, Connecticut, location served by highway but no transit or walkability to Boston's Seaport District, which is walkable and has mass transit. Other factors played a part, but both Connecticut and Massachusetts are high tax states and both offered incentives.

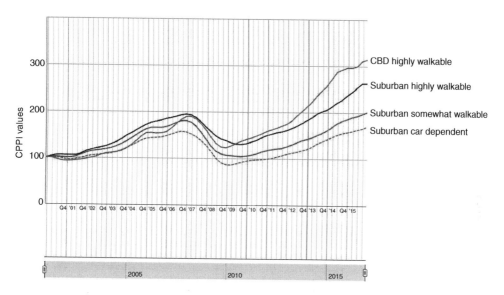

Figure 1.12 RCA and Walk Score® Commercial Property Price Index
Source: Real Capital Analytics; RCA & Walk Score® CPPI; *Moody's/RCA US National CPPI.

Another real estate indicator is the changes that have swept the retail side of the business. The mall is not dead (though there have been significant casualties) but has been transformed by place-making. Power centers and "big boxes" certainly continue but are no longer as rapid growing. As the baby boomers age and the millennials dominate, and given the growth of online purchasing, retail now has a far greater percentage of restaurants and less clothing or traditional department stores.

There is a unique place for the now thriving major 24/7 gateway cities[26] such as New York and San Francisco, but other cities, such as Raleigh-Durham and Austin, are benefiting from today's economy. There are still cities and metropolitan areas that are struggling such as Toledo, OH, or trying to find their economic drivers for the future. An interesting study by two Harvard post-doctorate students found that after decades of cities in the United States growing more alike, since 1980, they have been growing apart.[27] So as Richard Florida says, "you are where you live."[28] There are places, cities, suburbanism villages, that have the mix of jobs, features, transit and lifestyle that are flourishing, and others that are struggling. While some urban renewal projects including less than successful suburban style malls (such as in New Haven), today's redevelopment features much less auto based, much more mixed-use and pedestrian friendly active retail centers.

A separate study by the tech sponsored think tank Economic Innovation Group further found that the poorest cities essentially missed the economic recovery while the wealthier more knowledge-based economic regions grew strongly. This study further supports that the gap between rich and poor communities has widened, making urban redevelopment more important and more difficult in poorer cities,[29] and that another, also innovative measure of urban redevelopment, is the continued growth of the brownfield movement, the reuse of contaminated sites. A recent US EPA study measured the substantial growth and impact of the federal brownfields program, which is replicated to varying degrees in all fifty states. Recycling land, not just putting a fence around vacant former industrial properties, is a key

Table 1.2 EDR ScoreKeeper Brownfield Index

Market	Top 10 high growth markets (1st 6 months of 2016, Y on Y) (%)
Nashville, TN	17
Milwaukee, WI	13
San Antonio, TX	13
Raleigh, NC	12
Hartford, CT	10
Portland, OR	6
Oklahoma City, OK	6
Boston, MA	6
Orlando, FL	4
St. Louis, MO	3

Source: EDR Insight, ScoreKeeper model PPR market growth rates.

Note
EDR Insight's ScoreKeeper model tracks environmental due diligence activity (measured in terms of the volume of Phase I environmental site assessments) for the US market, regions, states and metros. Since due diligence is performed prior to a property transaction, Phase I ESA hot spots are a leading indicator of growing commercial real estate investment markets – much like the Architectural Billings Index is an economic indicator of future commercial real estate construction. As shown in the accompanying table, smaller secondary markets with strong growth profiles are seeing investor interest. Diane Crocker, EDR.

aspect of urban redevelopment. The above table is from EDR, a major data provider for environmental assessments and indicates a surprising list of cities that have seen more environmental studies for redevelopment of brownfields in the past year.

Urban redevelopment is always complex; there is no single smoking gun that totally captures the trend, just as there is no one silver bullet that results in revitalization. Clearly, urban trends are changing. There is more urban redevelopment occurring, some in center cities some in older suburbs, and each of the multiple aspects from waterfront design to real estate financing, contributes to this overall pattern.

Urbanophile case study by Rod Stevens

The following 2011 "urbanophile" blog post by Rod Stevens was triggered by a visit to Detroit, whose 2013 bankruptcy came to symbolize renewal failures, despite good intentions and even some good design. It also features a graphic that shows the "thirty-one" flavors of urban renewal.

> Aaron Renn's March 24 posting on "The Logic of Failure" and his reference to "silver bullet" solutions for redevelopment and revitalization reminded me of my visit to the "Creative Cities Summit" about revitalizing cities, three years ago this fall. The setting, timing and venue could not have been better, at least in terms of provoking thought about how to do things better.
>
> The setting was Detroit, the time was October, 2008, when the financial markets were crumbling, and the venue was Renaissance Center ("RenCen"), the Robocop-like mixed use center that is headquarters for General Motors. I flew in the night before and opened my door that morning to a newspaper lying in the corridor on which the top headline read "GM in Merger Talks with Chrysler." This was the beginning of the end as the auto industry had known it for the last 100 years, and those very same corporate managers were coming to work 30 floors above me.

This was my first trip to Detroit, so I decided to take a quick ride on the People Mover to see the city. With a station attached to RenCen, this automated system took me on a loop around the city, on elevated tracks 20 feet above street level, without my having to set foot on the street.

The first thing I was conscious of was that this was supposed to be rush hour and no one else was on the People Mover. For that matter, there were few people on the streets. Some cars streamed off the freeway and almost directly into the RenCen parking garages, but not many and even fewer people were out walking.

Most of the buildings between the stops also seemed to be empty. Most of them were of the same pre-WW II vintage and quality as those on North Michigan Avenue in Chicago and in Mid-town in Manhattan, but there was no one in them. It was like an old Star Trek or Twilight Zone episode in which something has happened and the population has disappeared.

I gradually became aware that many of the stops were at Sim-city like attractions – the kind you are allowed to build when your city gets to a certain size – such as the convention center, an arena, a baseball park, a football stadium, and a casino. Each of these must have taken hundreds of millions of dollars to build. I thought, "They've been spinning the roulette wheel, hoping to get the tourists and suburbanites back into the city." But what had the city fathers done for the residents themselves? Later I was to walk through Campus Martius, a center city park that people take considerable pride in, but even in the middle of the day it was largely empty. On the last day of my trip I walked up Woodward Avenue, the grand street at the center of the city that used to be the main place where people shopped. The buildings on one side were largely empty. The buildings on much of the other side were simply gone; some torn down for underground parking garages that were to be the new base of new office buildings to be built by private developers. These office buildings didn't materialize.

After this trip I began to compile a list of the "silver bullet" solutions of redevelopment projects that city leaders have put in place in various places across the U.S. over the last sixty years like those I saw in Detroit, and that I present here. Early in my career I prepared marketing and feasibility studies for these things, so I knew there would be a number of different kinds, but I was still surprised at their number when I stopped counting. Like Baskin Robbins, there are 31 flavors on the list, and it would be easy to add to it.

I have divided these into three kinds of projects: business, retail and tourism, and transportation. The bars show the decades they span, from the 1950s through the 2000s, with the earliest kinds of projects shown first. Expos, first on the list, actually started with the Crystal Palace and the Great Exhibition in England in 1851, which later inspired Chicago's "White City," but my time frame here starts after WW II, when American cities began to consciously redevelop themselves in the face of suburban competition. For each kind of project I have also included an example and the year that example opened. The examples were not always the first built, but they inspired others to follow. For example, the Ontario Science Centre came before the Exploratorium as a modern, hands-on science center, but it was the Exploratorium that most of the other centers in the U.S. looked to as an example. San Diego's Horton Plaza was not the first downtown mall, but it excited a lot of talk in the world of urban development.

Notice that the largest category is retail and tourism. If you really looked behind the rationale for most of these projects, you would find that most were in fact aimed at tourists or at suburban shoppers who had fled the city. The grand-daddy of all

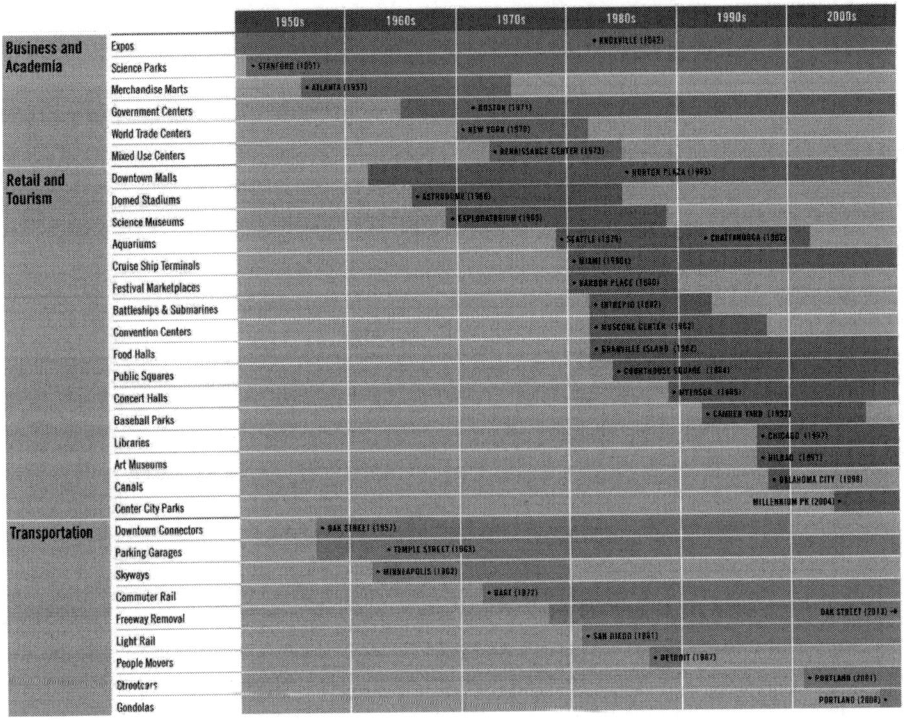

		1950s	1960s	1970s	1980s	1990s	2000s
Business and Academia	Expos				• KNOXVILLE (1982)		
	Science Parks	• STANFORD (1951)					
	Merchandise Marts		• ATLANTA (1957)				
	Government Centers			• BOSTON (1971)			
	World Trade Centers			• NEW YORK (1970)			
	Mixed Use Centers			• RENAISSANCE CENTER (1973)			
Retail and Tourism	Downtown Malls					• HORTON PLAZA (1995)	
	Domed Stadiums		• ASTRODOME (1965)				
	Science Museums		• EXPLORATORIUM (1969)				
	Aquariums				• SEATTLE (1976)	• CHATTANOOGA (1992)	
	Cruise Ship Terminals				• MIAMI (1980s)		
	Festival Marketplaces				• HARBOR PLACE (1980)		
	Battleships & Submarines				• INTREPID (1982)		
	Convention Centers				• MOSCONE CENTER (1982)		
	Food Halls				• GRANVILLE ISLAND (1982)		
	Public Squares					• COURTHOUSE SQUARE (1984)	
	Concert Halls					• INTERSON (1985)	
	Baseball Parks					• CAMDEN YARD (1992)	
	Libraries						• CHICAGO (1997)
	Art Museums						• BILBAO (1997)
	Canals						• OKLAHOMA CITY (1998)
	Center City Parks						MILLENNIUM PK (2004) •
Transportation	Downtown Connectors	• OAK STREET (1957)					
	Parking Garages		• TEMPLE STREET (1963)				
	Skyways		• MINNEAPOLIS (1962)				
	Commuter Rail			• BART (1972)			
	Freeway Removal						OAK STREET (2013) →
	Light Rail				• SAN DIEGO (1981)		
	People Movers				• DETROIT (1987)		
	Streetcars						• PORTLAND (2001)
	Gondolas						PORTLAND (2006) •

Figure 1.13 The thirty-one flavors of urban renewal

Source: Graphic by Carl Wohlt from an original chart and information by Rod Stevens/Spinnaker Strategies.

redevelopment projects is Ghirardelli Square, which remains vital to this day; although the upper floors have now gone condo for rich people who want to keep a place to stay in the city. Ghirardelli inspired the festival marketplaces of the 1980s, many built by the Rouse Company, and many of which are now struggling. These later morphed into the food halls inspired by Granville Island in Vancouver and the Pike Place Market in Seattle, and, more recently, the market halls or sheds for farmers' markets that have recently begun to show up. Notice the trend here for an ever-more-local clientele. Partly this is due to retail trends. When Ghirardelli first opened, it was filled with unusual boutiques selling clothing and glasses not found in the malls. Today you can buy these things at suburban "town centers," where chains like Crate and Barrel keep a good selection of wine glasses and linens.

There is almost a flavor-of-the-month approach for transportation as well, which really started with the downtown connector freeways aimed at whisking shoppers to ailing main streets. More and more cities are now tearing out these freeways and converting the space to parkland. What's more interesting is the evolution in rail, from heavy systems like BART and the DC Metro, to light rail in places like San Diego, to the current passion for street cars. Transportation is becoming lighter than air, and now there is even an urban gondola in Portland, with Vancouver planning a second on Burnaby Mountain. Years ago Disneyland had one of these, for frenetic visitors eager to punch all of their E tickets.

Notice how few kinds of business-related projects there have been. Science parks – which started with Stanford and the Research Triangle – have mostly been in the suburbs, but a few are in the city, such as Yale's Science Park, and more are on the drawing boards. Carnegie Mellon's Collaborative Innovation Center may be the best example of integrating academia, industry and the city, for here private sector tenants come together on a campus in the middle of a very urban city.

Notice just how briefly projects like Renaissance Center were popular. John Portman, an Atlanta architect, designed the most prominent of these, including not only Renaissance Center but the Hyatt Regency/Embarcadero complex in San Francisco (which is connected with sky bridges), Peachtree Center in Atlanta, and the Bonaventure Hotel in downtown L.A. At the time these were the wonder of their cities, and tourists came in to gaze upward at the atriums and light-bedecked elevators that moved through these. They almost all included office buildings, hotels, and mini-shopping malls, and almost never housing. Many of these were introverted, arrived at by car in special drop-off lanes, with the pedestrian entrances being hard to find. Few or no lobby windows faced out onto the street. At Embarcadero Center in San Francisco, the main level of pedestrian activity is one floor above the street and for about 20 years it had a thriving trade of office workers from nearby buildings eating and shopping there at noon. Now most of that lunchtime activity is out walking along the Bay, on the true Embarcadero, or eating in the Ferry Building next to it.

And then there are the truly wacky projects, which may or may not work in their own right. Projects like the automated people movers in Detroit, Miami, and Morgantown, West Virginia. The canal in Oklahoma City's Bricktown "entertainment" district. And the submarines and battleships, like such as a submarine in the prairie land of Muscogee, Oklahoma and the dreadnaught *Olympia* at Penn's Landing in Philadelphia. The *Olympia* ship may be headed to the scrap heap, for lack of support and visitation and Penn's Landing has struggled because of its isolation. Fish don't shop.

Why is it that these projects work in one place and not in others? And why is it that Portland has pioneered so many of these projects? I believe the answers are related, and having grown up in Portland, with a family that was involved in creating some of these solutions, I can offer some insight.

Aaron Renn uses the term "silver bullet," and that is exactly why many cities copy other cities' solutions: they hope these will magically solve their problems. But as Aaron points out, these other cities frequently fail to adequately define what problem they are trying to solve, and what their priorities are. The approach that works well in one city for one set of challenges will not work well for a different set.

But why has Portland been so successful? I believe there are three reasons: 1) crises and political turnover that opened the community up to questioning and new leadership; 2) a growing facility with problem definition and problem solving; and 3) the attraction of "outsiders" who joined the community and brought fresh new approaches and energy.

Without getting into too much detail, the political crises included a revolt and mobilization of the citizenry in the 1960s, when private interests tried to take over the public beaches. Never before had the legislature seen so many private citizens flood its conference rooms – and this led to other conservation measures like the bottle bill, land use planning and the Willamette River Greenway. This activism, growing at the same time as Vietnam War and Watergate, brought a new generation to power in the early 1970s, including Neil Goldschmidt, who pushed forward a light rail system when citizens revolted against more freeways. And the final event was a very, very deep recession in the early 1980s, when most of the major timber companies closed shop or left town, leaving behind a vacuum of power in which it was easier to make broad-based decisions. Oregon's growing environmental reputation and the easier entry into the circles of power drew in like-minded people from throughout the country, and some of these people helped push the city in new ways.

Most importantly for Portland, and perhaps for other cities, the community got better at problem solving, at not simply reaching for off-the-shelf solutions. In the 1970s, relatively strong retail, on the street downtown, led the community to reject a multi-block retail project connected by sky bridges that was proposed by Canadian developers. That first light rail line took care of a transportation need when citizens said no to a freeway that would have wiped out miles of neighborhoods. "Fareless Square," downtown, was a response to federal air pollution rules that made it tough to build new parking garages. The streetcar that opened in 2001 simply connected an already-strong downtown with Northwest Portland, a strong residential neighborhood that is the densest in the state. Portland has had its failures and misspent money – the Rouse project is now ailing and the extension of the transit mall has killed retail along its length – but its successes come because they are rooted in local needs.

Hopefully this trend is developing nationally. The failed Rouse project in Milwaukee, aimed at drawing tourists back into the city, is now re-oriented to more local shoppers, largely because Mayor John Norquist would not give it more subsidies. Money is flowing out of big downtown projects and into more neighborhood-based retail projects, like those sheds and squares for farmers' markets. And we are putting more of an emphasis on "productivity" projects, aimed at creating good places to work, and fewer on the "consumption" side, retail and housing. More cities are realizing that great places draw good talent, and that they need to focus on the work side if they are going to participate in the modern knowledge economy. Already we are seeing more collaboration between the city and the universities, and while much of this new development still takes place within the walls of the campus, in some places like downtown Phoenix, where the new Arizona State University campus has opened, the city and the university are one, without walls. It will be in leveraging the talents of our people, and our anchor institutions, that we do our best problem solving, and create the most interesting and durable of places.

To say that urban redevelopment has evolved is an understatement. Perhaps the most important change is that the top-down approach, regardless of how well intentioned or effective, no longer functions. Community development has morphed even beyond stakeholder involvement – the latter a term from the environmental movement. In the 2000s some worried that we had lost the ability to do large projects – but there are now many large redevelopment projects underway from Hudson Yards in Manhattan to Stapleton in Denver. It now takes long-term strategic and well-capitalized effort including concerted communications, not just meetings and attractive designs, but internet communications, expert consultants and public-private partnerships for large projects to succeed. Equally significant, there are now neighborhood redevelopments, often led by non-profit organizations such as SOBRO (South Bronx Overall Redevelopment Organization), CCLR (Center for Creative Land Recycling) and many others.

There is now a remarkable renaissance in many North American cities. The decline in industrial use has led to the opening up of opportunities, including waterfronts, for increased residential, recreational and commercial use. In an era when traditional suburban development has become difficult due to transportation costs, environmental concerns and most importantly market shifts led by millennials, there are significant urban redevelopment opportunities, with some outstanding examples to serve as models. The challenge is to provide a framework so that revitalization can be expedited, brought to more economically stressed areas and made more common, with greater emphasis on long-term sustainability.

Urban redevelopment today is extraordinarily complex, incorporating real estate economics, land use, community benefits, ecology, transportation, sustainability, place-making design, politics and a host of associated disciplines. There's also an array of regulatory and funding agencies, at federal, state and local levels, and often elaborate impact analyses and mitigation strategies. Development concerns such as: site analysis and acquisition, land re-use approvals, market analysis, financing, synchronizing remediation, redevelopment and liability protection, project organization and sequencing, design, and a host of regulatory and community reviews are all involved. Among the important strategies are: leadership roles and team-building, innovative financing tools including government programs, techniques such as charrettes, checklists and critical paths to aid information flow and support creative planning and design.

The history of urban redevelopment is just moving into a new phase, with market forces now moving in support. While there is no one magic silver bullet, there are a set of strategic pathways toward successful urban redevelopment.

Notes

1 Caro, Robert, *The Power Broker*, Alfred A. Knopf, 1974.
2 Ibid.
3 Jacobs, Jane, *The Death and Life of Great American Cities*, Random House, 1963.
4 *Kelo* v. *New London*, 545 U.S. 469 (2005).
5 Richmond, Peter, *Ballpark: Camden Yards and the Building of an American Dream*, Simon & Shuster, 1993.
6 https://nextcity.org/features/view/the-great-east-baltimore-raze-and-rebuild.
7 http://dailycaller.com/2015/05/14/most-baltimore-cops-are-minorities/.
8 www.baltimoresun.com/news/maryland/bs-md-housing-segregation-20151212-story.html.
9 Fulton, William, Governing, October 1, 2015, www.governing.com/authors/William-Fulton.html.
10 Heller, Gregory L.; Ed Bacon: *Planning, Politics, and the Building of Modern Philadelphia*, University of Pennsylvania Press, 2013.
11 Cahn, Amy Laura, "On Retiring Blight as Policy and Making Eastwick Whole," *Harvard Civil Law Civil Rights Journal*, 2013.
12 http://articles.philly.com/2015-10-17/entertainment/67591639_1_darby-creek-city-hall-residents.
13 Gallagher, Leigh, *The End of the Suburbs*, Penguin, 2013.
14 Leinberger, Christopher, "The Next Slum," *The Atlantic*, 2008.
15 NAHB, Quarterly National Housing Survey, January, 2015.
16 Kotkin, Joel, "The Geography of Aging, Why Millennials are Headed to the Suburbs," *New Geography*, 2013.
17 Sobel, Lee, Steven Bodzin and Ellen Greenberg, *Greyfields into Goldfields: Dead Malls Become Living Neighborhoods*, Congress for the New Urbanism, June, 2002 by Lee S. Sobel (Author), Steven Bodzin (Author), Ellen Greenberg (Contributor), & 2 more; 4 out of 5 stars 1 customer review.
18 Malizia, Emil and David A. Stebbins, *Urban Land Institute Magazine*, July, 2015.
19 Barrington-Leigh, Christopher and Adam Millard-Ball, "A Century of Sprawl in the United States," *Proceedings of the National Academy of Sciences of the United States of America*, June 2015.

20 Barrington-Leigh, Christopher and Adam Millard-Ball, "A Century of Sprawl in the United States," *Proceedings of the National Academy of Sciences of the United States of America*, June 2015, 112(27): 8244–8249. http://dx.doi.org/10.1073/pnas.1504033112.
21 Florida, Richard, *The Rise of the Creative Class*, Basic Books, 2002; *Rise of the Creative Class Revisited*, Basic Books, 2012; The New Urban Crisis, Basic Books, 2017.
22 Becker, G.S., *Human Capital*, University of Chicago Press, 1993.
23 Dunham-Jones, Ellen and June Williamson, *Retrofitting Suburbia, Urban Design Solutions for Redesigning Suburbs*, John Wiley & Sons, 2011.
24 http://jrc.sagepub.com/content/52/4/486.abstract Robert J. Samson.
25 Real Capital Analytics and WalkScore, http://news.theregistrysf.com/wp-content/uploads/2015/04/untitled.png2014.
26 Kelly, Hugh, "What Makes the 24-hour City Clock Tick?", *Premises*, Vol. 1, No. 1, New York University Schack Institute of Real Estate, 2011 and The 24-Hour City, Routledge, 2016.
27 Barrington-Leigh, Christopher and Adam Millard-Ball, "A Century of Sprawl in the United States," *Proceedings of the National Academy of Sciences of the United States of America*, June 2015.
28 Florida, Richard, *The Rise of the Creative Class*, Basic Books and *The Rise of the Creative Class Revisited*, 2012.
29 Schwarts, Nelson D., "Poorest Areas Have Missed Out on Boons of Recovery, Study Finds," *New York Times*, February 25, 2016.

2 Historic preservation

Barry Hersh

The preservation, conservation and protection of structures, landscapes or archaeological artifacts of historic and cultural significance are an inherent and important contributing factors to urban redevelopment. While 1950's urban renewal often wiped the slate clean, today's urban redevelopment preserves key buildings and historic enclaves and uses these to enhance the urban experience. While many urban redevelopments involve legally designated historic buildings and districts, others are less formal. Even if there are neither designated historic sites nor preservation funding, urban redevelopments today seek to have a context to remain part of the urban fabric. The reuse of older buildings and the echoing of building forms and details, whether or not technically official historic rehabilitation, often seek to retain the character, scale and authenticity that are important to communities and in the marketplace.

We preserve buildings and other features of the man-made environment because they make a better place, to tell the story of this place and carry on that history. There are three approaches to historic preservation. The first is historical, asking the building to tell the story, the history of its place. The second approach seeks to use the building and its surroundings to enrich the experience of the place. The cultural approach uses the building to contribute to the consolidation of cultural identity. Urban redevelopment can include any or all of these approaches, but the second, seeking to enrich the experience of a place, is most intrinsic and the most frequent element of designing a redevelopment project.

Every society seeks to preserve the man-made features that distinguish their culture, and North America is no exception. In the United States, the earliest historic preservation efforts started with local and state efforts to preserve George Washington's Headquarters State Historic Site in Newburgh, New York, and his residence in Mount Vernon, Virginia. Civil War and Revolutionary War battlefields were among early preservation efforts. The National Trust for Historic Preservation was established in 1949, and the 1966 National Historic Preservation Act (NHPA)[1] and later amendments significantly supported preservation of historical and archaeological sites by creating the National Register of Historic Places, the list of National Historic Landmarks, and the State Historic Preservation Offices. Among other things, the act requires federal agencies to evaluate the impact of all federally funded or permitted projects on historic properties through a process known as Section 106 Review.[2] Since 1976, the federal government has provided historic tax credits for the preservation of buildings and other historic features in accordance with standards promulgated by US Parks Department and administered by the IRS and State Historic Preservation Officers; numerous states also offer historic tax credits. Canada created what is now known as its Historic Sites and Monuments Board of Canada in 1919 and has designated historic sites and expanded historic preservation since then.[3]

Figure 2.1 Virginia City, Nevada, historic Main Street
Source: Wikipedia.

The ability of US municipalities to preserve historic structures and districts through their zoning or related ordinances was established by the landmark U.S. Supreme Court case preserving Grand Central Station.[4] This case, led by the New York Municipal Art Society supported by former First Lady Jacqueline Kennedy Onassis and others established that local historic regulations were not a governmental taking requiring just compensation, but rather a legal exercise of the government's power to enhance a community as long as owners continued to have some economic use of their property. Many communities, from New Orleans to San Francisco have adopted historic preservation ordinances, usually implemented by a historic preservation board of some sort, which often include architects, historians and others with specialized expertise. Many urban redevelopments that include historic structures or are in historic districts require approval from such historic boards.

There have been many outstanding examples of historic preservation contributing to the revitalization of cities, with Charleston, South Carolina, being a famous example. Starting in 1947 and continuing for the past forty years under distinguished historic preservationist and urban design advocate Mayor Joe Riley, Charleston has preserved and restored numerous parks, waterfront promenades as well as hundreds of buildings that have become a key element in the reinvigoration of this coastal community. Many citizen groups including the historic preservation society have played a key role in preserving Charleston's historic qualities.

Other famous examples include the preservation of the Old City of Quebec, Canada; Savannah, Georgia; Native American settlements such as pueblos throughout the Southwest; the Pioneer Square-Skid Row Historic District in Seattle, Washington. The Historic

Historic preservation 23

Tax Credit has proven a potent tool, providing a financial mechanism that supports restoration and preservation of historic buildings. Some municipalities have ordinances that prohibit or delay demolition of historic properties.

There are many examples of where historic rehabilitation has been a key aspect of an urban redevelopment project. In Ambler, Pennsylvania, the 1897 Boiler House was abandoned by its industrial owner in the Great Depression. Sixty years later, Summit Realty Advisors saw an adaptive reuse opportunity, a site with "good bones," transit-served location 30 minutes from Philadelphia and usable infrastructure. The site did require extensive remediation, including removal of hundreds of cubic yards of material contaminated with asbestos and other toxics.

The 48,000 square foot building, formerly a two-story plan, entailed excessive floor-to-floor heights for offices, so architects Heckendorn Shiles inserted structural elements to create three levels with clearstories bringing in more daylight while retaining a 140 foot tall iconic smokestack The building was awarded Leadership in Energy and Environmental Design (LEED) Platinum and includes an extensive geothermal heating system. Tenants, including the developer, high tech and healthcare companies have occupied this successful example of urban redevelopment by adaptive reuse.[5] This project has led to others, including the proposed Ambler Lofts project nearby.

Adaptive reuse is often a key element of urban redevelopment, turning old factories into lofts or office buildings, former banks into restaurants and former schools into senior housing, sometimes housing some of the former students. Many types of buildings and infrastructure can be adaptively reused, preserving an element of history while embracing growth and creating new opportunity.

Figure 2.2 Toronto Brickworks
Source: John Vertelli, Wikicommons.

It has been argued that adaptive reuse is significantly more sustainable than creating new structures from scratch. It keeps unnecessary waste out of landfills and limits unnecessary energy use from creating new materials; though the trade-off may be less energy consumption and other efficiencies possible in a new structure. A small specific example is an environmental consulting company, Enviro-Stewards of Elmira, Ontario, installing a "living wall" bio-filter that improves air quality inside a 100-year-old former furniture factory that features exposed brick wall and structural steel beams along with new energy efficient windows.[6]

There are several great examples of adaptive reuse in Toronto, from ongoing revitalization projects like Market Street to already finished transformations like the Toy Factory Lofts. There are hundreds of similar projects going on in different cities, but these Toronto examples show how fruitful adaptive reuse can be.

Examples of adaptive reuse in Toronto: Evergreen Brickworks[7]

Located at the heart of Toronto's massive ravine system, the Evergreen Brickworks is a series of buildings created to accommodate different brick-making techniques and levels of demand. The Brickworks were eventually abandoned, leaving behind empty shells in place of once busy factories and seriously contaminated soil.

By renovating these buildings, adding new, elevated structures (flooding is a major concern for most of the area) and creating a network of bridges between buildings, architects gave this old land new life. Its only new structure serves as Evergreen's head office, and the old factory has become a thriving community center hosting workshops, tours, festivals and other events.

Toy Factory Lofts

Built in the early 1900s, this building originally served as Irwin Toy Company Factory in an industrial zone west of downtown that has undergone a dramatic revitalization as Liberty Village.

Completed in 2008, this award-winning example of adaptive reuse preserved the factory's unique attributes and converted the space into a combination of office and live/work units. The revitalized building is also home to Balzac's Coffee and the Liberty Village BIA (Business Improvement Area).

North Toronto Station

Formerly known as Summerhill CPR (Canadian Pacific Railway) Station, this building first opened in 1916. In 1930, after struggles created by the Great Depression and the growth of Union Station, North Toronto Station closed its doors. Brewers' Retail moved in a year later, and the LCBO (Liquor Control Board of Ontario) started renting space there in 1940. For many years, they boarded up most of the space and let the building deteriorate. In fall 2000, revitalizing of the space began. It was restored, and a new entrance was built along with some retail stores underneath the railway bridge. In 2003, the building reopened, primarily featuring one of the city's largest liquor stores. This adaptive reuse helped preserve this building's iconic clock tower on Yonge Street and will be preserved for the foreseeable future.

Historic preservation provides verisimilitude and authenticity to urban redevelopment. It is often challenging to incorporate existing historic buildings into new development.

Figure 2.3 Toronto North Station
Source: Canada Rail.

Colonial era, federalist style structures tend to be small with low ceilings and tight spaces. Victorian era buildings, such as Queen Anne or Gothic Revival designs, may be larger buildings but have inefficient layouts and ornate features. Even twentieth-century modern buildings now over fifty years old and eligible for historic designation do not meet today's expectations in terms of energy efficiency, layout or amenities. Rockefeller Center is completely rehabilitating several of its landmark buildings including studios such as for the *Tonight Show with Jimmy Fallon*, improved glazing and energy efficiency and an updated Rainbow Room.

There are ways to incorporate older buildings. Retail today is as much about the experience and place as about goods; so many former mansions and estates have become places for dining and catering. Communities that retain historic buildings and landscape tend to attract all types of people: young and old, local shoppers and tourists. Historic waterfront redevelopments including Baltimore's Inner Harbor, Boston's Faneuil Hall and New York's South Street were a series of projects by the Rouse Organization in the 1970s and 1980s that helped define both historic preservation and urban redevelopment. Newer developments, including Georgetown in Washington, DC, and Portland's waterfront moved well beyond the formula, towards unique projects that capitalized on historic and waterfront characteristics. Even new urban developments try to capture some historic qualities and tie into the existing urban context to create a sense of place.

The renovation of former industrial buildings, especially multi-story mills, has been a key component of urban revitalization in many cities. Starting in the 1960s, artists started moving in to industrial buildings in Manhattan, south of Houston Street. Formerly referred to as "hells hundred acres" by firefighters because the industrial buildings lacked sprinkler systems and were often packed with boxes and other flammable materials. Many buildings were no longer functional and were underutilized. The infusion of artists, some from nearby historic Greenwich Village when it became pricey, became so strong that by 1982, New York City created the Loft Law,[8] which created a Loft Board and a process by which building owners and tenants could convert their buildings, bringing them into accordance with fire and health regulations. Over the next few years, SoHo became a chic and highly desired

26 B. Hersh

residential and retail neighborhood, with many buildings rehabilitated, some using historic tax credits. A new historic district was also created, but also with new construction on vacant lots and with less usable buildings. Many of the loft style apartments, featuring high ceilings, large windows and open floor plans, became highly desirable and very expensive.

SoHo also became the prototype for the pattern of artists and the gay community being pioneers, moving into a well located but run-down neighborhood with historic authenticity. Upscale boutiques and coffee shops soon followed, and then came real estate developers seeing an opportunity and young but affluent new residents. The range of former building types, from brick mill buildings throughout New England to cast iron structures to more recent industrial and school buildings were found usable for this type of reuse. Neighborhoods from Deep Ellum in Dallas to the River Arts District in Asheville, North Carolina, which features restoration of a former cotton mill, have been revitalized around the rehabilitation of historic industrial buildings. SoHo had relatively few residents (the original loft occupants were violating the zoning code), but as this pattern was followed in other communities, gentrification and displacement of existing residents became an issue, as discussed in Chapter 7.

A related, relatively new reuse concept for such buildings is the Urban Vertical Factory; the idea is bringing back multi-story buildings for at least partial use for new forms of manufacturing.[9] Nina Rappaport authored an exhibit and book that has brought forth this concept. There are many others, including Adam Friedman of Pratt Institute, who are strong advocates for retaining modern manufacturing uses and industrial zones in cities,

Figure 2.4 Denver Dry Goods Building
Source: Jonathan F.P. Rose Companies.

primarily to provide good jobs for immigrants and others without college degrees. "Makers" is a term for a new category of manufacturer, often using high tech and creative approaches ranging from three-dimensional printing to artisan metal-working, and are environmentally and economically amenable to urban areas, including the reuse of existing industrial buildings.

The Denver Dry Goods building was the successful redevelopment of an historic commercial building for mixed-use, including retail, by Jonathan F.P. Rose Company. A third generation real estate developer, Rose is well known for urban redevelopments that feature not only historic preservation but also affordable housing and sustainability.

Often retaining historic structures and quality is responsive to community aspirations, a desire to recall and improve upon history. Retaining key structures and design aesthetics can help build support for a redevelopment.

For an urban redeveloper, historic preservation requires special skills, architects and construction contractors who specialize in this highly technical area. Incorporating existing historic buildings into modern redevelopment often demands creative agility and can be costly. Finding materials and using older techniques that allow authentic restoration can be complicated. Each historic building is different, and if historic tax credits are involved, the SHPO (State Historic Preservation Officer) must approve all work. Starting with Columbia University in 1973, many major universities offer (mostly graduate) degrees in historic preservation, providing sophisticated education in the analysis, preservation techniques and policies of preserving historic structures and landscapes. The perennial PBS favorite television series *This Old House* is one of many sources of information for homeowners and craftspeople working in historic restoration.

Historic preservation is often a house by house effort as individual owners lovingly restore their own residences. Individual landmarks are frequently preserved by local

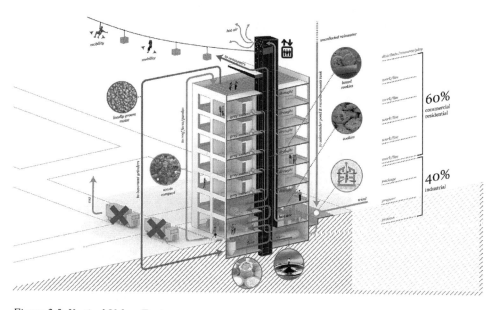

Figure 2.5 Vertical Urban Factory

Source: Image concept by Nina Rappaport and Natalie Jeremijenko. Illustration by Francis Waltersdorfer Vertical Urban Factory, Nina Rappaport.

volunteer historic societies, such as local museums. In terms of urban redevelopment, preservation is often about preserving not just one or two buildings but rather an entire district, while allowing redevelopment of properties that do not contribute to the historic nature of the community.

The move towards historic districts, preserving the feel of the community as articulated by Jane Jacobs and others, has become a key focus in the United States and Canada. New York City's first historic districts were Brooklyn Heights and Greenwich Village, and there are now more than 100 districts including over 27,000 buildings and 1,265 individual landmarks. Despite many redevelopment success stories, some real estate developers and others, including Harvard professor Edward Glaeser in *Triumph of the City*,[10] believe that there are now too many historic districts that are limiting growth and redevelopment. One of the more telling developer perspectives is that it may be best to redevelop just outside of an historic district, gaining the historic cachet, without all the regulations.

For urban redevelopment, the balance between old and new, between authenticity and modern amenities, between funky and sleek, is crucial. One perspective is that urban redevelopment can preserve industrial and what is called mid-century (twentieth) modern structures, that are often not as interesting to historic preservation advocates as, say, Federalist or Queen Anne style houses. Urban redevelopment can connect to more recent generations and help rebuild a community and connect to modern design as discussed in the next chapter.

Bethlehem Steel

Bethlehem Steel was once the second largest steel manufacturer in the United States dominating the relatively small (75,000 population) City of Bethlehem and the Lehigh Valley. Since the 1960s, the industrial giant abandoned one facility after another. One of the largest was in Lackawanna, in western New York State, which is now being redeveloped for industrial use and notably for Steelwinds, an extensive set of electric power generating windmills. In Bethlehem, the extravagant headquarters office building, constructed in 1972, is now long vacant. Its future has been uncertain since the main plant closed in 1995. The Bethlehem Steel Mountaintop campus had been acquired by the neighboring Lehigh University through a combination of gifts and state aid.

By 1999, there was an overall plan for the redevelopment of the 1,800 acre former steel mill itself[11] calling for the adaptive reuse of the steel plant, including its iconic blast furnaces. The project proceeded, albeit slower than initially hoped. Tim Fields, then EPA Assistant Administrator, called the plan a model of how RCRA (Resource Conservation and Recovery Act), the remediation program for industrial facilities, was intended to work.[12] The remediation was accomplished under Act 2 of the Pennsylvania Brownfields program that allowed relatively expeditious cleanup approval. Perhaps the least likely and most controversial aspect was the sale, approved by a one vote margin in the city council, to the Sands organization for a casino and eventually a hotel. The casino has certainly been a boon for the city and the rest of the parcel, leading to $900 million in infrastructure investments and 2,400 jobs being created. In a recent financial quarter, revenue was roughly $113 million; another $15 million came from Sands' other facilities on site. Pennsylvania requires casinos to pay a 55 percent tax on all revenue. Four percent of that goes directly to the gambling center's host community. Being in a zoned Tax Increment Financing (TIF) district (see Chapter 8 for a discussion of TIF) that revenue then subsidizes other infrastructure projects on the Bethlehem Steel site.[13]

"When we were looking for other cities that did a great job integrating a casino into their city we couldn't really find any," says Bethlehem Mayor John Callahan. "I decided that Bethlehem was going to be the city that did it right." Bethlehem was noted as a possible model for the redevelopment of Detroit in a 2013 *Washington Post* article that focused on creative adaptive reuse.[14]

Currently, Bethlehem Redevelopment Authority is setting aside $16 million for shoring up vacant structures that it will one day convert into a mix of residential and retail space. An interesting design feature is the proposed transformation of the Hoover-Mason Trestle (an internal rail line in the steel mill) into an elevated walkway, modeled on Manhattan's High Line as discussed in Chapter 9. Another imaginative but challenging historic aspect is the National Museum of Industrial History, proposed soon after the plant closed in 1995 and is scheduled to open in 2016. This new 20,000 square foot facility, built on the remediated plant site, features artifacts not only from Bethlehem Steel, but also through an affiliation with the Smithsonian Institute, industrial artifacts from around the United States. The museum will also benefit from its proximity to Lehigh University (noted for engineering) as well as Northampton Community College and the famed historic core of Bethlehem, located directly across the Lehigh River.

The Lehigh Valley region has seen a resurgence of light industrial development, such as Lehigh Valley Industrial Park, providing warehouse, distribution, office and light industrial spaces. The plan for the former Bethlehem plant retains CSX rail service and has targeted that market in the form of a Commerce Park, and now in an area east of the casino, a warehousing and distribution center is proposed. The largest new building proposed so far which has received planning approval is the Majestic Bethlehem Center's proposal for a 1.75 million square foot warehouse; eventually, it intends to build as much as eight million square feet of real estate.

Figure 2.6 Bethlehem SteelStacks Levitt Pavilion
Source: Bethlehem Redevelopment Authority.

Perhaps the most locally popular addition so far is SteelStacks, a 10 acre cultural center that uses the idle blast furnaces as a backdrop for public green space, a performance pavilion, an arts center and a new studio for the local PBS affiliate. BDA (Bethlehem Redevelopment Authority) has already spent $27 million on the project, money generated by Sands revenue. In addition, there is both the Steel Ice Center and the Steel Fitness Center nearby providing additional recreational opportunities,

The new arts center won a Pennsylvania AIA (American Institute of Architects) silver medal last year. Next door to it, SteelStacks' Levitt Pavilion is composed of a green space with appropriately steel construction features that lead up to the uniquely designed stage. Philadelphia-based WRT, which also designed Levitt Pavilion, received six different awards for the project. SteelStacks was a result of collaboration between the city and Sands, which sold 2.5 acres of the site to the city for one dollar.

Sands remains the most powerful partner the city's had to redevelop the area, and that's unlikely to change. The city and county are now exploring the idea of building a convention center, with Callahan calling the idea at a press conference, "a natural progression." It's no surprise that Sands, which runs convention facilities at its Las Vegas and Macau locations, believes such a project would be "right in our wheelhouse." As Sands and SteelStacks evolve, more cities are taking notice. Officials from other communities looking to build their own casinos recently checked out the site for ideas.

Notes

1 Public Law 89–665; 16 U.S.C. 470 *et seq.*
2 www.achp.gov/docs/CitizenGuide.pdf.
3 www.pc.gc.ca/clmhc-hsmbc/comm-board.aspx.
4 *Penn Central Transportation Co.* v. *New York City*, 438 U.S. 104 (1978).
5 www.bdcnetwork.com/power-plant-office-ambler-boiler-house-conversion#sthash.QqCnpl5T.dpufIn.
6 Building Design and Construction, November 20, 2015.
7 www.gardinergreenribbon.com/adaptive-reuse-toronto/.
8 New York City Multiple Dwelling Law, Article 7-C, 1982.
9 Rappaport, Nina, *Vertical Urban Factory*, Actar, 2016, based in part on earlier exhibit of same name.
10 Glaeser, Edward, *Triumph of the City*, Macmillan, 2012.
11 Wright, Andrew, "Recycling Steel," *Engineering News Record*, November 4, 1999.
12 Ibid.
13 www.citylab.com/work/2013/01/bethlehem-steels-redevelopment-winners-and-losers-public-private-partnerships/4394/.
14 www.washingtonpost.com/business/case-in-point-in-bethlehem-pa-a-road-map-for-detroit/2013/07/25/a744d070-f3dc-11e2-a2f1-a7acf9bd5d3a_story.html.

3 Urban design and city form in redevelopment

William Schacht

Introduction

Some have suggested that urban design is architecture writ large,[1] or that urban design is about the public realm. The Boston Redevelopment Agency (BRA), the sponsor of numerous highly acclaimed projects, defines urban design as "the art and science of shaping a city's public realm, collectively comprised of public spaces, the activities that occur within them and the buildings that frame them."[2] As an architect with direct experience working with BRA and honoring this basic definition, urban design today can be seen as a public-private partnership – focused on community improvement. Most large civic works now done in North America use some form of public-private partnership, which can be defined "cooperation among individuals and organizations in the public and private sectors for mutual benefit."[3] Such cooperation has both a policy dimension and an operational dimension, linked so that the participants contribute to the benefit of the broader community while promoting their own individual or organizational interests. Urban design now must include three defining transformative considerations: resiliency, sustainability and globalism – the pervasive reinventions required through the changes of climate, energy, technology and the world economy.

Resilience starts with the ability to recover strength quickly, featuring design that keeps occupants safe and secure – even in the face of catastrophes. Sustainability now deals with not just energy efficiency or designations, but the long-term viability of an urban design. Finally, globalism includes design that is open and accessible to all, including people from all over the world. It is now up to the urban designer to master the classical skills of the architect, engineer and artist, and to fully embrace and incorporate the new technical skills of aesthetically providing stability, safety, security, permanence, maintainability and renewability – in a knowledgeable and meaningful way.

The scope of urban design ranges in creations: the size and function as small as an urban vest pocket park, or as large as the urban districts such as Battery Park City and the new Hudson Yards in New York or "new towns" such as Reston, Virginia, and Stapleton, Colorado. While the practice of urban design has existed for as long as there have been cities building and rebuilding, redevelopment is now especially intertwined with urban design, involving not only the architecture of buildings and histories of place, but also as importantly, the spaces in between: streetscapes, urban plazas, infrastructure, linkages, landscape and the mix of uses.

These key factors now consistently emerge within the compositional considerations of any contemporary urban design project. The traditional concerns of orientation to sun, wind and weather are being altered due to these requirements. Classic palettes within urban

Figure 3.1 50 Hudson Yards designed by KPF
Source: Photograph by T. Lawrence Wheatman.

design variously utilizing color, texture, shade and shadow, calculations of hourly or seasonal nuance and poetic level change (for example, L'Enfant's Washington DC[4] and Old Montreal's historic port waterfront) now need to be regarded in new ways. At the same time, too many twentieth century urban places developed seem arrested, to have intentionally disregarded these concerns or superseded them in a metal and glass scale less subservience to a higher rate of passage.

Design for urban redevelopment must address the basic issues of safety, ownership, construction, accommodation, transportation and civil engineering. Historically, a traditional, regionally distinct expression grows through the appropriate incorporation of local building materials such as clay, wood, brick or stone. Traditions were born and slowly developed. Now, glass, metals, stone and materials from global sources dominate increasingly denser vertical solutions in North America and around the world, and there are some urban thinkers that strongly advocate for greater density.[5] In order to reduce the carbon footprint in contemporary construction, LEED and other sustainability standards reward the use of local materials. Far beyond the goals of locality, the special considerations of the art and science of sustainability can simply be defined as being a "design philosophy that seeks to maximize the quality of the built environment, while minimizing or eliminating the negative impact to the natural environment."[6] Similarly, the UN Brundtland World Commission has proclaimed that we will attain sustainability and a positive legacy when we redevelop the built urban environment (by urban design) by "meeting the needs of the present without compromising the ability of future generations to meet their own needs."[7]

The place-making of an urban design expresses both priority and hierarchy. Government centers are typically urban designs of great importance, answering the need for places of civic observation and interaction, celebration, dedication, homage and ceremony – as developed, for instance, by Edmund Bacon at Philadelphia's Independence Mall.[8] Today, we further incorporate the new expressions of place and event designation with the possibilities of social media communications and the responsibilities of resiliency, sustainability and security. New cultural manifestations continue to merge into the expressions of our civilization, whether with the new inferences of security bollards and the intrusions of cameras and check-points or the positive technological considerations of vast new climatically controlled centers for commerce, entertainment and/or civic discourse. These contemporary elements – transformed or newly created – emerge today at a far greater pace certainly than during the pre-machine, evolutionary age of urban design.

What is seen is often controlled to a large extent by the unseen. Oral and written standards and practices have been with us for a very long time. These are generationally carried forward – occasionally modified through individual or collective insight. Codes and regulations – since the early twentieth century in the US and Canada (and since Hammurabi in history) – define possibilities. Urban design guidelines have now become a more common control for urban places. This form of palette guideline control has recently been used to excellent advantage in several urban centers: the design guidelines of Seattle and those of Vancouver, British Columbia (discussed here) come to mind. Looking at urban redevelopment includes brief thumbnail analyses of several recent North American case studies through interviews with the authors and peer evaluations of urban redevelopments of several large and small cities and the illustrative case studies through this book.

There are new requirements in terms of federal, state (or province) or local restrictions, including environmental impact and resilience, public access – including the disabled and

– since 9/11 – regard for safety and security. The approval process and compliance are art forms in and of themselves, but new possibilities also occur with the celebration of resiliency, sustainability, clean energy, new infrastructure and transportation and other technologies.

Urban design is, in another sense, a culminating expression of civic accomplishment, a triumph of fiscal commitment and creation, a place of ribbon cutting, of photo-ops and perhaps overblown statements of individual accomplishments, especially when historic or cultural landmarks are involved. Urban design may also create the spaces that function as a new civic front door or sometimes as an exterior entry reception hall, an urban corridor, an entertainment venue – places of civic pride, a lively mix of uses, opportunity or even exhilarated activity. An urban designer can draw a great new place at an enlarged scale. Urban design plays a key role in all urban infrastructures: train stations, transit facilities, bridges, tunnels, and even parking structures – all add to the diversities and richness of the city environment. While a project may be civically needed and supported, it must also be carefully planned and integrated into the community. The very essence of an urban design project is to engage the community and use ideas that arise from public discussion, reaching consensus and transforming expressions of declaration into a new civic design.

Urban design uses a communal and visual vocabulary in its mission to gain consensus and respond to the community needs required of an ever urbanizing, shifting society. Design leaders are given the charge to create a grand synthesis of numerous required programmatic elements – and they are at all scales and – seemingly – infinite. To make a profound piece of music, we need a unique composer and a masterful conductor. To make a unique urban place, our orchestra must also be superb, capable and thoroughly rehearsed to construct and properly execute the designer's vision. To complete a major urban design, we also need to find the right moments, when the financial and political resources necessary are compliant and available to produce a successful project.

Incredibly, there remain those at the highest levels of authority who espouse a nonbelief in the notion of a "Master Plan!" As if a reliance on market forces, intuition or "wise" zoning policies will fill the bill! Folly! Look no further than the efforts of a Robert Moses or a Rob Ford to see that there will always be the basic mandatory need and necessity for a true urban design plan with concomitant flexibilities and comprehensive inclusions. The economic efficiencies of the planned and coordinated needs of infrastructure serving a general planned arrangement of civic functions and fiscal responsibility demands that such allocations are pre-planned. Reliance on chance and occasioned subjectivities will only significantly waste urban resources. A coordinated study of the myriad of competing requirements is the only way to effectively save the community treasure. A comprehensive urban design plan, by whatever name – addressing the community goals and objectives – is the tool that the responsible urban designer utilizes to provide the necessary basic design framework.

There are numerous physical, constructional and cost constraints in urban design, as well as requirements to find community consensus that all work together to generate a meaningful design framework. This also – to be validated – inevitably requires compromise. Architects, engineers and urban designers must not only be warriors for aesthetic quality, but also must be clever strategists to follow through on an urban design master plan over what is most likely be a long-term, multiple year engagement. This is a time of incredible urban transformation, globally and microscopically, to not only define and articulate community priorities, but also to utilize in a specific location the available technical, social, cultural, environmental and financial new order to create exceptional new urban places. In a time when the definition of an urban center is undergoing a basic scrutiny of relevance and a reevaluation of resilient need, new urban redevelopment thinking and manifestations are also critically needed.

It is the urban designers' role to direct the myriad forces that move an urban project, to help build momentum and to incorporate a range of stakeholder voices. It is our good fortune that the growth of cities in the U.S. has also led to a greater appreciation of significant urban place-making and a renewed recognition by many, including major institutions such as the United Nations and the Roman Catholic Church, of the importance of urban environmental resilience and sustainability. Because of this added dimension of recognition, urban design and redevelopment must also deal with the issues of social equity such as affordable housing and the neighborhood distribution of resources, such as libraries and schools.

Urban design process

An urban design is complex, but it must be led by a singular leadership team and given appropriate authority. While a design may spring from an individual, there are always the prerequisites of the team. Strong leadership and good communications are always the hallmarks of a successful design team. A unique combination of specialized consultancy firms will be required for each project and will vary with the project composition according to the nature of the redevelopment, the member specialization and the interfaces with the developer, the uses and especially the required interfaces with government and community representatives. Generally, a long-term urban design "master" plan is created to guide future phases of development, while allowing for a reiterative process and flexibility as field conditions, market conditions and policies appear and shift.

Urban design team members generally include:

Design professionals:

- Urban Designer
- Urban Planner
- Architect
- Landscape Architect
- Transportation Engineer/Planner
- Civil Engineer
- Structural Engineer
- Mechanical and Electrical Engineer
- Environmental Engineer

Specialized design consultants:

- Sustainability/LEED/Energy
- Mechanical specialists
- Hydraulics and Hydrologist
- Materials: stone, concrete, metals, glass
- Wind (tunnel)
- Climatologist
- Lighting designer

Related professionals:

- Attorneys: real estate, land use, finance, environmental
- Insurance
- Community Liaison
- Public Relations
- Marketing and Real Estate Brokerage
- Environmental and insurance experts

Parameters

An urban design project must define its parameters by a careful consideration of the existing conditions of:

1. *Physical analysis*: Considerations apply regarding natural features such as site mapping, topography, geology, water courses, ground water and vegetation. Climate, wind, viewshed and orientation to the sun are all important fundamental environmental site features to be thoroughly regarded as well. Man-made factors such as infrastructure, including roads and transportation system, utilities including water supply, sanitary sewer, storm systems and telecommunications all need to be appraised, designed and integrated.

2. *Community factors*: Urban redevelopment inherently means change to an existing community. Understanding in considerable detail what that community is, what the residents want and assuring true community engagement is essential in the planning and design process. This certainly includes, but goes well beyond, collecting and analyzing socio-economic data and trends regarding families, housing stock, institutions, and health and economic conditions. Community issues must be addressed as early in the design process as possible and resolved as much as possible. The desire to preserve all things as they are, sometimes referred to as NIMBY (Not In My Backyard), to prevent all change – even if thoroughly developed arguments present undeniable evidence that these changes bring real community benefit. Community identity comes with strong physical familiarity. Strange and change are typically shunned by those who would "protect" the nature of the place.

Urban redevelopment means neighborhood change, even in historic districts, and change is always hard. The urban designer can guide the redevelopment by working with an ingrained sense of the communal consensus in terms – for example – of residential living patterns and allocations of dwelling unit size, and in terms of shape, character and intent of how institutions, community facilities and public spaces are perceived and used. Communication and real interaction is essential; the designer cannot be validated nor effective working in a vacuum.

One can trace how centers, cities and civilizations morph over time. Even today's North American urban communities tend to shift from one culture to another. The Mexican-American neighborhood of Los Angeles continues to expand; the town of Dearborn, Michigan, has transformed from German and English to now a predominance of recent immigrants from the Middle East; and Flushing, Queens County, New York, has totally changed from Italian-American to Chinese and Korean-American.

With these shifts come different patterns of lifestyles and special needs. Different religious worship practices require different or adaptable spatial relationships. Different types of cuisine require different retail food services and restaurant demands. Different social interactions demand different assembly needs and holiday parade routes, for example. Music, poetry and art and their expressions are accommodated in a number of ways and various types of venues. Parks that once needed baseball fields and bocce courts, then more basketball hoops, now feature soccer fields. Traditions and expressions of legacy vary from culture to culture and from generation to generation, often connected to demands for different types of public spaces and amenities.

3. *Financial feasibility*: Ultimately, it is the responsibility of the designers to have an understanding of what can and cannot be done – given current financial resources and anticipated markets. This is obviously an extremely important restraint. As discussed in Chapter 8, Real Estate, there are often many private and public financial participants in an urban redevelopment project, and the design must inform and work with that framework. There is a need for flexibility to respond to constantly changing market demands as well as community expectations.

4. *Zoning and land use regulation*: A key aspect of urban design is to design within existing zoning criteria or request what may be difficult changes to the existing land use and zoning regulations. These exist in almost every community throughout the United States and Canada (with Houston being a special case). Zoning is normally based upon an official, formally adopted set of physical and functional controls within a municipal master plan, with allowed land uses being a key element. These rules are often complex: starting with building heights and setbacks, floor area ratio and allowed uses, but also expanding into nuances such as plaza transfer of development air rights (how super-tall residential buildings are allowed in New York City) to planned unit development in smaller communities. Zoning and other land use regulations play a major role in determining the shape and nature of urban redevelopment, affecting large and small areas, impacting aesthetics, schedules and costs, and often requiring public hearings and community input.

Today's urban redevelopment projects often involve what often becomes a complex set of negotiations, in which the urban designer is a key participant. In the past, and still for smaller projects, these negotiations are based upon the requirements of a site plan or other land use permit – often a set of conditions that must be followed. For larger projects, and those with complications, such as brownfields or transit facilities, these deliberations not only involve more agencies, but transform the municipality or community agency from a regulator to a participant. Especially when there are mutual financial commitments, these become public-private developments, often memorialized in extensive contracts. Often larger redevelopments have their own detailed urban design master plan and unique district, separate and distinct from the overall community master plan that is the basis for municipal-wide land use and zoning regulation.

Urban redevelopments are often part of various types of jurisdictional districts beyond normal zoning. There may also be a specially created "planned development district" whose regulations are separately negotiated for a large, specific project. All or part of a redevelopment may be within a designated historic district, as noted in Chapter 2. There may also be a special tax district, such as for Tax Increment Financing, as discussed in Chapter 8. This may allow advantageous municipal bond financing. All of these designations may have advantages, but they may also impact the project approval process, schedule and ultimately the urban design.

Smaller projects, such as gas stations, dry cleaners and other individual sites may fall within existing zoning, though their configuration and use of an existing structure may improve eligibility for zoning board approval of variances. Each site is unique, and every municipality has their own zoning ordinance and community master plan – but many local political officials are supportive of redevelopment efforts that meet community expectations.

38 W. Schacht

Miami-Dade County is among the jurisdictions that have adopted a new type of "form-based" zoning, entitled Miami21.[9] Developed by locally based but world renowned new urbanism architects Andres Duane and Elizabeth Zyber-Platek, the zoning uses transects – a visual typology of place – to describe what a community, especially the public realm, will look like. The zoning then creates the rules with a flexibility to accommodate that vision. An alternative to conventional land use regulation, form-based zoning[10] in a sense starts at the end, with what a place should look like and then fills in the regulations to accomplish that goal, as compared to traditional zoning which starts with rules (use, heights, FAR, setbacks). These and other innovative zoning approaches are aimed at providing urban designers more flexibility such as mixed uses and varied street widths, to encourage both urban and suburban development. The following is an illustrative transect showing how different zones of development are portrayed.

Many states and cities now also require some form of Environmental Impact Statement, an analysis of how every aspect of a community may be affected by the project: from air quality, to traffic, sewer capacity, storm water runoff, skyline, housing availability, as well as historic and even archaeological impacts. These are generally required for large projects, and those that receive federal, state or local financing. It should also be noted that all projects must conform to U.S. federal requirements and the regulations of the American Disabilities Act.[11]

5. *Urban design guidelines*: In many larger and more sophisticated communities, such as Boston, Vancouver and San Francisco, there are – in addition to zoning and other traditional land use regulation – urban design guidelines, usually specific to a district or a large site. Design guidelines became popular after the success of Battery Park City in New York, an early (late 1970s–1980s) adopter of both design and sustainability regulations. For such large urban redevelopment districts, there is a guidebook of minimum standards and details produced to literally guide or regulate the design for each individual site so as to be a consistent component within the overall plan. There may also be a Request for Proposal (RFP) from either developers or their architects or a separate design competition for architects. The guidebook and/or RFP are in addition to the basic land use, massing heights and square foot areas established in the early overall Urban Design Master Plan.

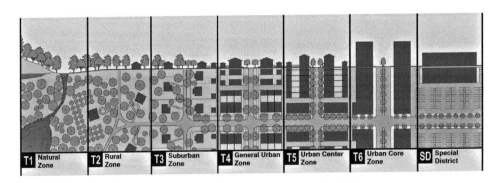

Figure 3.2 Transect. To systemize the analysis and coding of traditional patterns, a prototypical American rural-to-urban transect has been divided into six transect zones, or T-zones, for application on zoning maps. Standards were written for the first transect-based codes, eventually to become the SmartCode, which was released in 2003 by Duany Plater-Zyberk & Company

Source: Center for Applied Transect Studies.

Individual teams may be asked to comply with a solution that successfully addresses such requirements as:

- Entry locations and designs
- Base and superstructure materials and window glass area percentages related to facade total area
- Window sill height, type, subdivision and operation
- Individual or group building configuration interrelationships such as base cornice line characteristics, horizontal dimension limits without setbacks, etc.
- Structural bulk constraints to adhere to a properly scaled project development
- Maximum-minimum building dimensions and distance of property line setbacks
- Landscaping bed locations and acceptable planting materials
- Paving material options and patterning requirements
- Signage location restrictions, sizes and typeface requirements

Technology and tools for urban design

Technology is playing an increasingly critical role in the urban design process. In the twenty-first-century digital age, technology supports and is vital to cities, with new technology tools ranging from Geographic Information Systems (GIS), to advanced aerial photography, to what is called "big data." All informs investment and design decision making across the globe. The developments of digital media and information technology in Silicon Valley have impacted urban lifestyles – and urban design.

1. *Satellites and drones*: Technology initially created for advanced commercial and military aerial photography, other large-scale ecological and sustainability measurement, communications and for many other purposes is used. Global positioning systems have largely replaced and improved traditional land surveying and are used widely to measure man-made structures and natural features such as topography, water courses and vegetation. These increasingly used tools provide extensive data and images that are extraordinarily useful in the urban design.

2. *Design and visualization*: Designers today have a wealth of tools to help them draw and visualize their architectural and urban design projects; computer aided design is now the norm, replacing hand drawing except for conceptual sketches. Common computer proprietary design software is utilized to render extremely accurate and uniformly high quality custom plans, sections and elevations. Graphic images can be very efficiently and rapidly produced – in either two or three dimensions. Virtual reality electronic imagery is commonly utilized on most typical mid-scale and large-scale projects. Early stage concept design communications with the design team – including the owners and contractors – allows all general and detail aspects of the architectural, engineering and urban design to be reviewed at the time of its in-office creation-using Building Information Modeling (BIM). Even connecting to marketing site websites to show the potential end user experiential fact similes are now commonly available to close real estate transactions. Perspective views or walk-throughs of a furnished and finished place or a serial sequence of spaces can convey on command any aspect of the project for immediate team review. The capability of designers to visualize and present their concepts and electronically virtually walk clients and others through the proposed structures and public spaces has revolutionized the design process. BIM systems extend the process of generation and management of digital representations of physical and functional characteristics of places. These systems allow design

and construction teams to communicate in real time, with three-dimensional information, and to far more effectively coordinate with entire construction teams.

3. *GIS*: As the new cartography, urban planners and designers use Geographic Information Systems to attach a wide range of data to locations for analyses and presentations, ranging from locating contamination hot spots to showing demographic changes over time. GIS mapping can also be tied to design programs used by architects and engineers.

4. *Energy management systems*: Contemporary buildings and spaces have energy demanding systems including the heating, ventilation and air conditioning (HVAC), integrated fans and thermostats, lights and light sensors, and computers and elevators linked so as to monitor, interact and enable effective management of energy use.

5. *Real estate data systems*: There are a host of new, web-based tools used by real estate and design professionals in North America. These include market information such as provided for the single-family home, commercial market data, and office leasing and plan data.

6. *Big data*: This refers to systems that measure, sometimes in real time, everything from traffic and storm water flows, to shopper preferences and walking patterns. These help inform the urban designer, providing real information that may want to be reflected in design features and places.

7. *Security systems*: Detectors are given key locations on buildings and other prominent locations and have become design elements to integrate. Entry and exit control gates/road guards/pedestrian turnstiles and metal detection systems now provide normal, mandatory security control design elements that the urban designer needs to accommodate to visually co-exist with the more time honored and expected urban landscape features.

8. *Construction technology*: Exponentially higher urban densities have increasingly altered the nature and capacities of urban design public space throughout the twentieth century and continuing on into the twenty-first century. The increase in capability and capacity through the use of sophisticated steel and carbon fiber (structurally); concrete, polycarbonates, carbon fiber, titanium and aluminum (weight reduction); glass fiber and elevator and escalator technology (intercommunications and vertical transport); and internet/IT for supplanting public telephone booth locations with individual and group intercommunications sitting and standing space requirements.

The first mandate: safe, secure and resilient

It is the responsibility of those planning an urban redevelopment to assure – to the greatest extent possible – the safety and security of current and future users. Urban districts are formed and use a variety of approaches to provide safety and security. Of course, there were defendable walls built around the cities of ancient times. Today's defensible lines tend to be more subtle but with an incredible overlay of security and safety systems. From crime prevention to early warning systems to our national defenses, there are networks of safety and security to be considered throughout our cities – and in the districts and places within those cities. Herein lays the basic tenet of resiliency.

Catastrophic events are neither predictable nor totally possible to plan for and can often be most effectively dealt with at the level of urban design. We are seeing an increasingly frequent need to defend our cities utilizing urban design to better control acts of nature and the anti-social acts of man. We are now learning the lessons of the Netherlands and of Venice, Italy, and the realities of climate change and sea level rise become more apparent every year. After hurricanes Katrina and Sandy, as well as 9/11, the quality of an urban design is gauged significantly by the standards and requirements used in measuring resiliency. Several regions of the United States are still evaluating and beginning to implement numerous ways to rebuild wiser and stronger, and in some cases new knowledge and approaches have been applied and succeeded. We have seen, however, limited improved physical resiliency evidenced in places such as in urban neighborhoods from New York's Rockaways to low-lying communities in South Carolina and Florida.

Today's urban designers must anticipate and address protection from natural and manmade catastrophic events. As noted, waterfront developments must be protected against storms and flooding, and there are both government regulations and professional guidelines now in place. Certain critical decisions, physical manifestations and the organization of a security and a safety system are all standard requirements. In ocean, lake and riverine coastal areas, safety from storms, flooding and any exceptional high water levels is critical. Since recent major storms, the Federal Emergency Management Administration – much criticized after Katrina – has been issuing new flood maps, planning guidelines and revising insurance programs. At a minimum, key electric, heating, cooling and telecommunications equipment must be raised or otherwise protected. New and rehabilitated buildings are required to be raised above flood plain elevations. Twenty-four/seven microgrid systems are required to keep mission critical functions (fire, police, public ambulance, etc.) in operation when the predominant power and light systems have been incapacitated.

At the scale of urban design, it is appropriate to plan in some detail for the possible restriction of access points, emergency vehicle proximity and accessibility, etc. Vehicular access control (bollards, gates, pavement deterrents, etc.) and security guard stations and building entry ID controls (cameras, turnstiles, booths, desk consoles, etc.) are typically required as a part of new project standards. Commonly considered new project elements now include individual citizen computer chip intelligence, drone surveillance and a growing array of systems designed to monitor and hopefully prevent destructive acts of man as well as of nature.

For populations who live in urban redevelopments, the potential need for retreat is a key planning consideration. From congested urban areas, this requires planning of sufficient evacuation routes, coordinated transportation modes, collection and transfer points, and wide portals to safer locations.

Emergency centers need to be considered in new urban designs for communities vulnerable to natural disaster. The designs of Buckminster Fuller, best known for his creation of the geodesic dome, were among those prescient in planning for resilience from disasters. Many public facilities including schools, stadiums, commercial malls and other facilities – some with vast enclosed dome or retractable roof systems – can be used to provide storm protection, anticipating results far more successful than those experienced after Hurricane Katrina at the Superdome – the enclosed structure of New Orleans.

According to verifiable climatological information on heat and humidity level changes being generated – certain significant regional and urban areas may actually become detrimental to human habitation within this century. While urban areas have always used shade from trees and buildings to cool public spaces, today we can go much further in terms of site planning towards reducing the urban heat island effect in cities, encouraging use of solar and wind renewable energy.

There is no doubt that water is the precious commodity of the future. Comprehensive planning must take into account the capacities of existing sources to sustain any new design requirements. Capturing precipitation and recycling existing supplies by design will be required to provide acceptable equations of supply and demand. Maintaining landscaping and cleaning practices will need to be scrutinized to balance with other municipal needs.

As urban buildings grow ever higher and severe storms more frequent, wind considerations are an increasingly early and important aspect of the design process. Over exposed vulnerability coming from storm events with winds of 150–200 mph demand intense, detailed study. Wind tunnel tests and specialized engineering studies are required to evaluate design, especially of large and tall buildings. In more normal weather, consideration needs to be given to comfort in planning public spaces and amenities such as sidewalk cafes.

Current urban fire and emergency codes are constantly being evaluated regarding new technologies and provisions for protection and access. New urban places must have a comprehensive vehicle and water access plan. This is especially critical where the typical street grid is interrupted. It is vital that fire and police departments are fully brought into the design process so that citizens can be fully protected. Enforcement of building, fire and environmental codes is generally quite stringent in the United States and Canada, but there have been exceptions which put the public at risk.

California and some other parts of the United States have mandated requirements in their building code that improve the ability of structures to withstand earthquakes.

The urban design plan

The Design Plan, whether or not it is called a "Master Plan," may primarily be an informed arrangement of structures in an urban landscape; it most typically incorporates the entire process of assessment, planning, design and implementation. This is in service to resolve not simply architecture and historic preservation of buildings, but to engage necessary improvements of infrastructure, to create functional urban forms with capabilities that meet new and traditional human activity and to integrate contemporary regulations and functional requirements.

The boundary lines of an urban redevelopment are legally represented in contract and survey and are required to be clear to all. However, the impacts of a specific project will often go beyond those boundaries, and concerns will overlap with the broader existing community. Often easements must be negotiated to give appropriate access to the surrounding property owners and the public, but consideration for neighbors and community may be far more extensive.

As with an architectural project scope, there are distinctly defined design phases in an urban design project. Initially, a conceptual design plan is developed showing the broad outline and image. Once accepted by the design team, the concept is followed by a schematic design – providing more functional analysis – including costs – at a larger scale. Design development – with increasing specificity and detail – including detailed cost information and value engineering (a quality-cost options study) – is concluded within the various sectors of the plan. The design then is finalized after all the necessary public meetings and approvals have been accomplished. Actual construction documents are generally prepared on a building or facility basis, utilizing the entire urban redevelopment plan as a key plan.

Goals and objectives: The planning and design of an urban redevelopment must start with determining the project's goals and objectives. What are the most important needs to be met and from whose perspective. A private developer certainly wants financial return and must meet a market demand, but there are also community objectives in terms of housing, amenities and community character. Successful urban developments – whether private, public, non-profit or – very likely today – a combination, start by working collaboratively with the community to determine the overall goals and objectives that can then be translated into the program. There is an inherent mutual mandate between the designer and the developer, between financial, aesthetic and community goals. By identifying the key elements that must ultimately be in the plan, and by clarifying communication among the stakeholders, a consensus can be reached to achieve goals in a timely and efficient manner.

Program: In order to design a project, there needs to be a functional program based upon the goals and objectives of the project. This program is expressed in terms of a hierarchy, including the size and type of spaces that need to be provided. Lengths, widths, heights, square and cubic footages: all are unique interactive dimensions and capacities determined to best provide a desired interaction and character. Working with a congruent infrastructural framework, an urban design becomes defined by and supportive of those structures.

There is a budget and real estate feasibility aspect to almost all redevelopments and the design must make sense financially. The mix of residential/commercial/institutional/industrial spaces – to the extent programmed – must meet market requirements. Parks and other public amenities need to respond to community needs. The costs and expenses of the project, as well as the sources and uses of funds, as discussed in Chapter 8, must be considered as part of the design structure. There are iconic structures – often but not always supported by public funding such as major transportation stations and performance venues – that have seemingly limitless budgets. Inevitably, there are budgetary limits, however, and the sometimes unseen hand of the market place and practical need to which the designer must respond.

Review process: Urban design is a reiterative process; plans are made, changed and remade. The goals, program and ensuing plan and ultimate scope inevitably go through a formal and informal review process. Plans will be revised and updated to meet market, community, construction and other demands, in what is often an arduous process. The support and agreement that can be achieved in the program and planning phases will greatly assist in the community and governmental review process.

Urban fabric and infrastructure: Urban redevelopment design is generally built upon existing infrastructure including a range of utilities, site conditions and a transportation network. The redevelopment design must bring often inflexible systems up to current and future standards. There needs to be an overall monitoring system that keeps all sub-systems functioning. Even the most mundane of systems, such as street cleaning, trash pickup and equipment storage, need to be designed, and have constant monitoring and daily site maintenance. Some of the larger urban redevelopments, such as Hudson Yards in NYC, are modeled after the prototypes developed for Orlando, Florida's Disneyworld – with an invisible, mostly underground, trash handling system.

In terms of the infrastructure of transportation, the entire range of modes and operational requirements must be considered as discussed in Chapter 4, urban design and redevelopment is intrinsically tied directly to transportation infrastructure. We were led at that time by the pioneering urban designer Victor Gruen, whose planning and urban design work reflected that: "transportation planning is considered an integral part of the redevelopment effort, and improvements in the mass transportation system are significant elements of the entire planning approach."[12] There needs to be 24/7 public-private use of urban areas, and there must also be continuous vehicular and pedestrian access. This is true even at the extreme ends of use intensity: the most intense public use traffic as well as with the least frequently utilized vehicular service traffic easements.

Such a diverse array of transportation requirements must be accommodated by careful planning and engineering. These modes – each with its special design requirements – include:

Over-the-road tractor trailers: These are often required to service a new urban design site, and they need to be provided with very wide dimensional maneuvering areas and constricted loading/unloading dock areas. Access times are generally required to be off hours – often in very late evening hours. The last quarter mile is often the hardest part of a 1,000 mile journey.

Vans and pickup trucks: They require greater frequency and both dock and service door access.

Buses and jitney bus vehicles: These will have required passenger loading/unloading locations that comply with other lighter vehicle circulation. Lighter small capacity vehicles will access and distribute within larger open area plazas.

Subways and light rail transit: If below grade level subway access is possible to the site, this becomes a vitally important connection and gathering/distribution point. Special care in delineating the various types of movement within this station point must acknowledge cross path patterns, milling, queuing and gathering areas and pathways with access to retail kiosks, public toilet facilities, ticket dispensing and information central locations.

Taxi and livery vehicles: While queuing space must be allocated, the possibility exists that – as with the automobile – future access will be restricted or not accommodated within the various urban precincts. This has certainly been the norm in most European urban cores for quite some time.

Private automobile and parking: The automobile may be considered significantly antithetical to urban redevelopment, while parking demands often define a suburban design. Private automobiles, however, must be accommodated to some extent in even the most urban locations. If sufficient pedestrian and public transit is provided, there is less need for parking, while limited parking may be relegated to the periphery. Some central cities have taxed and all but banned autos, while it is also the case that large parking facilities are located on real estate too valuable for this function. As private vehicle technology changes, the nature of parking requirements will also change.

Bicycle: American cities are finally discovering what Europe has known for decades: bicycles provide better, faster and more economical transportation within the urban area. Park/storage/access areas need to be carefully planned and designed, along with plans for expansion. This is especially true for bicycle rental programs – as this long available mode of transportation is projected to grow substantially in the future.

All of these transportation modes are undergoing a constant reappraisal of hierarchy of use prioritization and access ability and emphasis. And with the development of driverless, automatically located and guided, renewably powered vehicles – the land use requirements for rights of way (especially high speed) and certainly private vehicle parking garages will be greatly reduced. Because of this, therefore, the availability and capacity of the urban areas served by these functions will become increasingly redeveloped for other uses.

An important element of urban design is the urban block grid hierarchy. Urban redevelopment has, as noted in the discussion of urban renewal and in some other modern cases, created "superblocks." There are cases – such as large former industrial properties – where this approach works well, but typically urban developments have more often tied into and utilized the existing or nearby street hierarchy. Human scale and a typically traditional smaller urban block – allowing alternative paths – is typically preferable, perhaps alternating in conjunction with a more limited intersperse of larger blocks.

Quite logically, the transportation and utility networks follow, parallel and overlap one another to similarly access all of the ongoing urban points of function. It is critical that these systems are in communication and coordination with one another. Contemporary systems are designed to simultaneously function without literally disturbing the paving surface and the subsoil buried utilities. Prefabricated removable panels surface a utilities trench that cleanly provides a continuously accessible channel for water, sewer, power and cable/communications rights of way. Repairs and replacements are done without subsoil or traffic disruption.

The three-dimensional relationships of all systems, the proximity of elements and layers must be accurately locatable on a master grid map that allows for all systems to be accessible for maintenance and emergency needs.

Construction: The urban design team plays a key role in managing the physical construction of the project, working with selected construction firms. Just as "all politics is local," so too all urban redevelopment construction is local. When the urban locality is not of sufficient size to have specialized or major local construction management firms to negotiate the highest order of required construction, then more regionally capable construction management firms will come in. Even then, the prudent design and construction organizations will associate with the smaller local organizations to access the critical local methodologies and recommended practices to identify the local materials, efficiencies and local lessons learned over the years. There are numerous risks associated with project construction; accidents, delays and cost over-runs among them. The design team, working with the construction managers, needs an ongoing proactive relationship as well as well-drafted contracts to minimize such risks. In addition, the actual construction project needs to provide a sustainability template to utilize the most sustainable local materials and methods of construction.

Spatial flexibility and adaptability: New technologies may demand the substantial rearrangement of programmatic functional workspace use. High technology firms now have need for far fewer enclosed private offices, requiring rather a variety of shared open working spaces and meeting rooms. The average square footage needed for each office employee is now half of what it was a decade ago. Flexibility for the inclusion of equipment consolidation versus expansion, and the changing basic character of the work space (natural light, ceiling height, building configuration, etc.) all spell the wisdom of the programmatic provision for increasingly large, open and adaptable spaces.

Similarly, exterior open space must be designed for various uses by various use groups at varying times of the day and/or night. Times Square, for instance, needs to serve variously as an outdoor international greeting great room, a civic/entertainment stage, an outdoor assembly hall for civic honor and a nightly reception forecourt to an innumerable number of cultural venues. Interlaced within great urban places and their linear extensions is the need for casual strolling, exercise, walking and other health related activities.

Metrics: A beneficial new urban design component – integrated into any urban redevelopment/urbanscape (as opposed to the creation of entirely new comprehensive urban construction) – must answer several questions regarding its new and maximized reasons for existence:

1 Does this urban design provide benefit to a comprehensive range of urban users?
2 Does this urban design reinforce or beneficially alter existing patterns?
3 Is civic opportunity afforded?
4 Are we regarding a civically developed program?

5 Will this urban design enterprise exemplify great clarity and leadership (and expertise in support roles)? How is this measured?
6 Has public-private resolution (philanthropy, private functional roles, civic voices and input) been fully explored?
7 Does this urban design provide new, unique or additional urban capabilities or options? Is it contributive of new and/or expanded capabilities?
8 Is there a regard and respect for historic context. Any example of recent/current urban design must certainly fit as one integrated piece into that of a much larger existing urban fabric puzzle. Exceedingly complex, any such addition and series of required connections will undoubtedly run into a certain degree of NIMBY reaction.

But regarding the consensus of the great civic majority, a new design must be given a chance to exist, to be used multiple times, appraised and adjudged.

Urban design form

August Heckscher succinctly observed that the classic vision of urban form was "of man rooted in a particular environment, given personality and cohesiveness of character by his relationship to an external world that both shaped him and reflected what he was."[13] He noted that our challenge of the twenty-first century is a more complex assignment, noting

> we have at least the glimmering of a new vision: man becoming largely independent of environmental factors (best equipped through the newest resiliency standards), transcending place, superior to old bounds and limits – sensitive to an extraordinary number of impulses, and in command of unprecedented knowledge.

So we have our challenges set before us within his presciently embracing overview.

And so, urban design form most rightfully must begin with the people. Here we must always keep in mind the words of Percy Johnson-Marshall – an honored Royal Institute of British Architects urban designer and urban planner – who states:

> Planners and architects, as the chief designers of our physical environment, should never forget that the most important thing about a city or town is the people. The city is essentially a stage for diverse human activity, and a physical environment reflects, often in a subtle way, the character and qualities of its inhabitants. The urban designer, in endeavoring to project a vision that is not just his own subjective idea of tomorrow, but is something that will provide for the varied physical and emotional needs of all the citizens, needs some kind of brief in terms of a human specification, not perhaps in terms of "what do people want?" as they tend to like what they know.[14]

Rather, the responsibility of the urban designer is to show through creative excellence how the unimaginable can be attainable reality – through new ways encouraging and convincing the community of these new real opportunities.

Overall civic form transformations will range from a peripheral border or edge enhancement to those of new primary civic focus and broad civic utility. It remains the case that there is no substitute for master architectural design authorship capabilities, even if distributed over many generations and individuals – as at Piazza San Marco in Venice or with L'Enfant's District of Columbia. Approaches taken by early twenty-first-century "starchitects" such as by Zaha Hadid, who found transformative ways to integrate and express the urban infrastructure; or by Lord Norman Foster or Renzo Piano, who elaborate upon the urban grid; all in contrast to Frank Gehry, whose emphasis is on the focal monument. All are modern approaches to the issues of historic integration and contemporary urban design.

Is urban design form perceived relatively statically as a pedestrian, or at what rates of movement if as a passenger/driver – in one of the various modes of conveyances previously discussed? Is the experience perceived as a singular event, or more commonly as a dynamic series of experiences moving to and through this place? The excitement of accessing civic participation within an urban place is in the sharing of the exhilaration of arrival, and sharing that attainment with the excellence, pride of place, and the generosity of that acknowledged and shared accomplishment.

Among the primary formal concerns of the urban design team are those considerations listed below:

Defining city spatial form: Typical "historically evolved" North American cities have a horizontal hierarchy – from a high-density center outward to a surrounding suburban sprawl beyond the principal city boundaries, first identified by Park and Burgess in 1925.[15] Each zone has its distinct and identifiable characteristics. Homer Hoyt first proposed the more economics based urban sector theory[16] of city form. How residents actually perceive their own city was first well-articulated by Kevin Lynch in *Image of the City*, 1960.[17] As the automobile impacted urban form, models such as the variegated network were developed for cities such as Los Angeles and Houston. Christopher Alexander, developed an early example of data driven analyses, observing how cities are works of man, not nature and that "A City is Not a Tree" in 1977.[18] Much more recently, 2011, Edward Glaeser's *Triumph of the City*[19] is a strong defense of density and high-rise structures and critical of a perceived over-reaching of historic district preservation.

All of these noteworthy efforts, and many more, analyze the nature of city form, how neighborhoods have a mutually distinct, identifiable and supportive internal system of standards. This would include the nature of the density and heights of the buildings, the character and breadths of the streets and sidewalks, street furniture standards, traffic support characteristics, and graphics and traffic control standards. The nature of each district, the use of a grid or other street configurations, and the mix of uses are all part of the city form. Of course, as the city expands or contracts, these zone boundaries – always somewhat diffused – will shift somewhat.

It is imperative that the urban design team properly adjudge the existing character of the urban form and the site, and that the design falls within the appropriate context of that existing character or – if the area is to be significantly upgraded – to appropriately express that next urban step of character, density and/or design complexity. At the time of project analysis, the expertise of the urban designer needs to address the design opportunity with a deep regard for the rich design vocabulary provided by history.

Historic references regarding urban design form – and later in this chapter urban design elements – can be excellently illustrated by the work of Bernard Rudofsky in his book *Streets for People*. The photographic images following in this section to illustrate examples of urban design form and elements emanates from Rudofsky.[20]

Project composition: Once there is an approval of the site authorization and scope and content of the functional program, the responsibilities to the city are turned over to the master urban designer. Anticipation visits the urban designer able to inhabit this rarely offered key role – the opportunity to bring physical reality to those often multigenerational hopes, dreams and responsibilities programmed with and entrusted by the city fathers and participating project specific clients. It is the responsibility of the urban designer to offer a full range of concepts and feasible alternative solutions for the full consideration of the community.

Landmark, history and fabric: The design team must appraise the given design site in the context of the overall composition of the surrounding urban area of influence and how this new addition/modification can maximize its urban contribution. How does it connect/respond to the urban transit network to which it will contribute? Are there historic properties and cultural contexts to be preserved and enhanced? Is there a need for additional open space within these particular precincts? Most urban design projects today connect to the urban fabric, often continuing, extending or creating a street grid. Many follow author Jane Jacobs' preference for short lively streets, with shops, restaurants, streetscape and amenities and with nodes that become part of place-making.

Site orientation: Each site certainly has its specific context to be carefully considered by the design team related to its historical civic position, its cultural responsibilities and its physical relationships to climate, sun and views. Similarly, the design team must regard the site relationships to the centers of the surrounding urban area, and the perspectives by which to perceive these off- and on-site relationships. A juxtaposing of these existing centers to the new design will provide the context for the natural elements of orientation and connection with which to be explored.

Is this a site that can welcome the provision of significant areas of natural features? Will we be held to very subtle contacts or a very limited palette? In most cases, the case for maximizing nature in its fullest expression must be made – as usually we find that the site is being asked to infill an area already limited to accessible open space, and demanding additional density.

Scale: The sheer lineal dimensions of the site – incorporating both open and structurally enclosed spaces – require careful consideration. An open space on the scale of St. Peters or Ground Zero (coincidentally close in dimension) will yield a civic place where facial recognition is barely possible from end to end. Is this desirable, or do we want a more district-neighborhood/interstitial friendly space to gather or celebrate and acknowledge our neighbors.

Functional relationships: Each redevelopment design asks that certain primary functional interrelationships be addressed relating to physical or characterization proximity. These relationships are primarily addressed as structures to provide certain functions of certain general (or specific) size and scope. What is less specifically programmed or addressed are the characteristics or "needs" of the supporting open spaces or the interrelated connections between individual or grouped structures. Here the design plays a key role: creating new and unique places – and playing with those site orientation elements noted above – for the enjoyment of and the participation by the public-private citizenry given access. Here the concept of "node and pathway" comes into play. Within an urban design, certain functional focal points or "nodes" are orchestrated to interplay within a certain sequence or series of sequences. These experientially sequenced focal points are reinforced by the "pathway" or "pathways" that interconnect them. Each segment may have its own character or be a serial or sequential route of a commonality of urban design elements (paving pattern, bollard design, stepping sequence, street furniture character, etc.).

Boundaries and edges: The success of the insertion of a new place or district into the urban landscape depends to a significant degree on how well the edges of this new construction blend/merge and integrate with the surrounding urban fabric.

Urban design and city form in redevelopment 49

The designer must carefully consider these edges and how they resolve themselves on the macro level. New functional structures themselves often form the edges of a new redevelopment. The boundaries – wisely counter-positioned and formed with structural elements – can be the major functioning performer role in bringing this interface together or, in contrast, to keep district edges separate.

Where this contrast of old and new is most sensitively perceived and required is at the points of penetration via streets and/or open areas. Here pavement materials and patterns play a subtle but important role. Today's place-makers are well advised to use richer, more humane surfaces and paving materials than the omnipresent pavements of asphalt or concrete. Such actual interfaces between old and new require carefully considered configurations, textures and colors. Edge definers such as bollards, steps, low walls and planting strips will play major roles in bringing this edgework together. Also critical will be those places of passage: the streets and sidewalks. Here, the differentiation is more of an announcement, an entrance-exit vocabulary set and possibly a special structure such as a gateway, an arch or perhaps entry pylons.

Within the 360 degrees of a new circumferential urban design interface, there will be a multitude of such interactions and a broad range of decisions – both design and experiential. An excellent example of urban interface development through time comes to us by looking to that quintessentially American city: Chicago. And an excellent and highly recommended resource reference of American 360 degree urban development comes to us through the publication *Chicago: Growth of a Metropolis*, a book published by the University of Chicago press. Here, within these pages, we literally see a 360 degree study of Chicago and its urban relationships to its magnificent lakefront amenity – in "revealing" panoramic photographic documentation – from the same viewpoint – of the Chicago of 1858 (pre-"Chicago Fire"), 1913 ("Second City"), 1937 and 1969.[21] Any student of the American city would want to plead with the University of Chicago for an immediate 2016 panoramic addendum to and redistribution of this magnificent tome.

Some of these decisions regarding interface to existing buildings, complexes or districts address some assumptions regarding the projected civic life of those surroundings. Certain elements of this 360 degree edge will, like the new insertions themselves, require a prioritized, hierarchical regard. In a word, certain elements of the surrounding fabric will be held in greater relative importance based on functional interface, civic regard or projected life.

Perspective and sight lines: The perceptions of the pedestrian user are the key to the success of any urban place. The design team must demonstrate acute awareness to the cognizance of the different citizens as they transverse or pause within the place to be created. What are the various perceptions likely to be encountered within the site? How does one enter the site on foot or disembarking from an auto, bicycle or other transportation mode? Each has its priorities and limitations regarding sightlines and through rights of way. One must give way and space to those who pause (tourists) or are asked to pause. It is a necessity that the designer know or anticipate the capacities of those who will use the space. Times Square, for example, is jammed into a maelstrom of interferences and interruptions (welcomed and appropriate), while many midtown Manhattan plazas are underutilized, forlorn and failed.

Ascent/descent: One of the early determinants of the disposition of the physical elements of an emerging new place is the vertical on-site inter-relational requirements and capabilities. While many (perhaps most) new urban design sites need to resolve several vertical connection issues, some may offer astounding design opportunities to "go from here to there" in vertical resolutions through the means of the use of ramps, steps, staircases, walls or slopes to engage these geophysical vertical variations. Here, we must be especially cognizant of the catalogue of requirements of the American Disabilities Act (ADA)[22] and acknowledge the responsibilities (legal as well as moral) of the designer to provide a completely interconnected, accessible pathway through the place for the wheelchair bound or those with limited limb functional capability.

Hierarchy: The designer and the community client should find initial pre-design consensus regarding any desired prominences to be expressed regarding individual structures, interrelated structures or open areas within the site to be considered for recognition, commemoration or celebration. In the case of open spaces, the transitional access routes within (and without) the site are critical to a proper regard of these hierarchical requirements.

Focus: If a new urban place is adjudged to require the creation of a certain focus within the composition, the design team must analyze this requirement as to how to best create this focus, be it by formal positioning of the elements, embellishments within the focal element(s) themselves, or by the addition of enhanced texture, light and/or color.

Entry/exit and pedestrian access adequacy: Just as in every building or occupied structure, there is an identifiable entrance or entrances and exit or exits, so too it is with an urban place. Typically, multiple entry and exit points to an urban place are usually related to a surrounding transportation grid of streets, often extending that very grid through the site in a similar or enhanced manner. Importantly, the creation of a new urban place gives the opportunity to alter that typically repetitive pattern. Where that pattern alters, therein lays a new opportunity of special configuration and place emphasis, a special place of respite and a special place to be and to reassess. Access to and from these central places are best clearly marked architecturally, given a prominence and designed to serve as dramatic passage and/or as a quieter subtle point of retreat. The "front doors," for example, to the historic features around Independence National Historic Park in Philadelphia are important as they call to attention the main central historic buildings as an integral part of the urban space experience.

As discussed earlier in this chapter regarding safe and secure considerations within any new urban design, any successful urban space must provide clear and observably comfortable entry/exit points. At a minimum, two clear and distinct pathways need to be provided to the surrounding adjacent districts for its users. The actual as well as the psychologically reassuring provision of multiple and ample exit points is one of the primary tenets of a safe and secure place.

Oscar Newman carefully studied the interactions between the new inhabitants of a newly designed complex and the user inhabitants of the existing surrounding communities. His reason was to define and enhance interaction safety, comfort and well-being. In his book *Defensible Space* – referred to later in this chapter as a case study – Newman states in his Urban Locale section that:

> If particular urban areas, streets, or paths are recognized as being safe, adjoining areas benefit from this safety in a real sense and also by association.
>
> It is possible to increase the safety of areas by positioning their public zones and entries so that they face on areas which, for a variety of reasons, are considered safe. Certain sections and arteries of a city have come to be recognized as being safe – by the nature of the activities located there; by the quality of formal patrolling; by the number of users and extent of their felt responsibility; and by the responsibility assumed by employees of bordering institutions and establishments. The areas most

usually identified as safe are heavily trafficked public streets and arteries are combining both intense vehicular and pedestrian movement; commercial retailing areas during shopping hours; institutional areas; and government offices.

Natural light, shade and shadow: Responsible design must need to take into account what proposed structures will impose on an urban place. Depending on the location, the sun may be a very welcome or a very unwelcome essential element in design decisions. What Stockholm welcomes, Dubai must carefully ward against. As a significant urban design principle, the renowned architect Moshe Safdie noted, significant and important major cities around the world have established mandated access to natural sunlight, for office structures in Germany, and for residential structures in China.[23] It is the view of the author that – with ever increasing urban density and orientation restrictions – mandatory minimal natural light accessibility should be required as an international standard.

As a "natural" design tool, the sun and its concomitant "natural light" potentialities are among the most powerful (and free). The enhanced texture of a properly angled wash of sunlight upon a rich and patterned masonry wall or the enhanced vibrancy of a spectrally engineered color/form accent dancing in the sunlight – these are the delights that can be brought to the urban participant, if the urban designer is aware, generous and responsible to his or her craft.

A vitally important consideration regarding composition and user impact regards the sun and how it strikes/bathes the site. Certain functions will be deemphasized in shade while others will be highlighted literally by their prominence in the sun. As the time of day moves from early morning until late at night, the mood and character of an urban place will naturally change. Here also is where artificial lighting systems take over in the evening hours. Windows dark to the observer during the day become highlighted tableaus. Glowing, day-lit facades turn dark at night. So too must the design team regard evening lighting composition reversals to again enhance prominence, create a nocturnal contrast or recede into a placid backdrop. Points of intense light will draw a prominence not afforded during the day – a factor that must be considered when rendering functional use controls.

Juxtaposition/superimposition/transition: The user of an urban space will experience a shifting relationship between the site elements as he or she moves through the urban place. Certain special places and buildings off the major place will gain or lose emphasis over time. Some important visual elements may only first totally appear during the course of the participant's progression through the place. This progression and arrested movement can be emphatically and uniquely used by the sensitively aware urban design team to heighten many of one's urban experiences. The elements of surprise and of altered emotions become great tools in the wise urban designer's toolkit.

An urban place may typically have more than one use and character. The design must acknowledge these various places and utilize a variety of techniques to emphasize or ease these changes.

Design elements

Human scale: At the scale of a unique or individual site vocabulary, there are certain subsystems that – with expert design consideration – can certainly give the participant an added sense of place, wonder and curiosity. But, as most learned urban designers profess, scale always begins as the measure of comparison – to the community participant – the human participant, and the human scale of man, woman and child. As Paul Zucker initiates us to this awareness, the "correlations of the principle elements that confine the square (read urban design) is based on the focal point of all architecture and city planning: the constant awareness of the human scale."[24] We start all urban design with our knowledge of and relationships to human scale. We measure with, for and in comparison to the human scale. We measure in view perception and angle, in ease of access acceptability, in reachability, in the ability to communicate, and in the relationships to common architectural accommodation of door, window, floor, steps, masonry unit size and other subconscious comparisons to the familiar. As the noted urban planner Hans Blumenfeld has expressed, this simple interconnection to the human scale and experience: "The presence of elements of familiar size, such as bricks or steps, give scale."[25] Steps are particularly important because of their three-dimensional character and their association with movement in space. Scale related design elements include:

Street walls and floorscape: An urban design composition is most typically derived by the composition/location and character of the street walls that contain the site functions of human activity brought to the site. The street walls are the primary composers and creators of the urban space. The wall heights, setbacks and inter-structural relationships of those walls are primary and predispose the further "levels" of wall refinements: the placement, proportion, repetition and emphasis of wall openings such as covered arcades, windows, doorways, entry points and gateway passages, as well other functional and/or ornamental accents. So too are the typical depth setbacks within the street wall facades of these elements and the three-dimensional "style" applicative of the framing of these wall elements. Hereto we may also be excited to discover true and unique sculptural elements embedded within the walls to delight or perplex us. As with similar floorscape concerns, street walls are not only designed to be primarily involved with regarding dimensional and compositional issues, but also regarding key issues of color, texture and finish as discussed following. The street walls and the floorscapes of an urban design are the primary elements that are the responsible holders to engage the basic issues and resolution of character and history, harmony and contrast.

Street furniture: Those elements that give a place its special character are the special monuments or commemorations, the fountains, the street lighting, the benches, the directional graphic signage, the drinking fountains, the vendor kiosks – all need to be an ensemble of a singularly related vocabulary – of harmonious elements (harmonious range of colors/tones, same metal or stone finishes, same functional profiles, same script, etc.).

"Fixtures of the plaza": Scale and activity generators to the urban design include vendors, street performers, guides, salespersons and kiosks, beggars and many other kinetic "fixtures of the plaza." These are the various community individuals who want and deserve to utilize the public realm to conduct business. Zoning, nuisance laws, and law enforcement monitoring and presence are often used to assure that the quality of a public space is reasonably and conductively maintained. While we treasure a freedom of movement, speech and individual entrepreneurship, an urban design needs to provide for the provision of these elements, while maintaining a certain sense of social control – sufficiently reassuring so that the public feels welcome and secure.

Color, texture and pattern: Certain local building traditions/cultures have initiated and in some cases maintained a certain color palette or expectation. Baltimore's red brick is an example. Whether it is the certain terracotta tone of the tile roofs and balconies, the vivid or muted tones of the rendered plaster walls or the special color set by the locally quarried stone, these strongly influencing hues do set a subtle subconscious tone. Similarly, the incredible variations that can be achieved in the finished surfaces of stone and brick masonry, especially when highlighted by the sun, can impart a delightful and varying accent to a place. These textures, of course, can be varied and patterned to emphasis these effects. The use of motifs and repetitions in paving patterns and building facade elements can be a powerful visual enhancement to the enjoyment of the public.

Graphic signage and way-finding: A system of signs or symbols to guide people appropriately through large urban places is often necessary and appropriate. At a minimum, street or pathway signage may be desired. It may well be the case that the existing citywide system could impose a less than desired visual result, and a new special district standard should be incorporated. When this is done in conjunction with the overall design composition, a complimentary unity may occur – in some successful cases elevating these routine functions to that of art.

Lighting: Contemporary nighttime lighting offers many options to enhance evening enjoyment, as well as to provide necessary guidance and safety. Lighting becomes a very important design element and, when used well, can engage the realm of that of a true art. Lighting design can have extraordinary power, transforming a place into a unique, nighttime attraction.

Nature: Even within the most urban of spaces, nature is omnipresent. Redevelopments are often close to or a part of waterfronts, parks or other natural features. We may embrace many of these offerings – even the most modest and unintended – and plan for nature. Those tiny cracks you may see opening in the joinery of a wall or the pavement of a streetscape will widen deepen and eventually collapse even stone, brick or metal. Even in the densest urban areas, birds will carry seed and insects to our design, in time giving flower to new trees and unremittingly altering all features. Landscape, basophilic and natural features are best knowledgeably anticipated, planned for and incorporated, rather than resisted.

Sun, with its daily and seasonal cycles, and wind, with its range of comforting assurance through threatening (and obviously seriously destructive) possibilities, is a set of constant presences that reconnects us to primal forces. And depending on the local climatological characteristics of a site – rain and snow and ice may become formidable factors, difficult to accommodate positively. Conversely, the occasions of heat and drought may significantly call for design accommodations such as awnings or other shade producing/water protecting elements. Local climate is a major serious design consideration, and is to be seriously regarded as a resource by the capable urban designer.

Life cycle, energy and sustainability considerations: Have the responsibilities of providing for a public-private place included adequate resources for maintenance, repair and replacement life cycle costs? Over the course of an anticipated multigenerational lifetime, these life cycle costs will often far exceed the original construction costs and attention. Here is where value engineering becomes an important determinant. If, for example, all elements of a design plan were specified to be among the most sustainable, there may indeed be a premium added to the initial construction cost. Comparative life cycle costs of employing energy saving and other green features, however, have come to be on a par, as Energy Star, LEED and other certification programs become standard. If, on the other hand, the least expensive materials were utilized, these materials would require more operational, maintenance and replacement cost over the reasonable lifetime of the place. These costs may well prove to be far more expensive, as well as far more disruptive to the ongoing utility and convenience of the place.

More speculative private or public-private ventures become, therefore, more problematic. The lesson here is that – for a significant redevelopment project – the importance of it as a civic improvement begs for the adequate budget and quality materials route. Better to insure a quality adequate, comprehensive budget and gain the advantages of a fuller life.

Public performance, art and sculpture: Public and private spaces can be greatly enhanced when providing the opportunity for public performance and demonstration. Gathering spaces must always be considered and positioned as an integral part of the public place. As importantly, cultural values need to be allowed expression through the careful inclusion and placement of important art and sculpture. Permanently placed pieces or a program of changing urban art in specified locations can give a place not only added significance through recognition, but added stature by celebrating the interactions between art and community. Historic or contemporary artists thus become an active participant in the social web of the place.

A public community spirit: If a majority of these elements and considerations are successfully incorporated and met over time with civic memory, then successful Urban Design has occurred – justifying permanence in the civic urban memory. A unique palette of elements will have been brought together in a special way, and society will respond to it – be it with exuberance, pride, passivity, obliviousness or less. Great success will either crown such a creation as one of art and spirit or one of simple utilitarian acceptance. Cities require both as focus or backdrop and this for reasons of program or contrast, as well as concerns of budget and resilience. Is it a special place to enjoy, to fully recommend to guests and visitors as something worthy of time and attention – as something comparable to those accomplishments of neighboring communities or enterprises? The place will signal an expansion of accommodation – acknowledging in some way the expansion of the surrounding community urban design capabilities for gathering and commemorating or for celebrating special civic/historic events. Enlightened developers have determined that special places and quality design will in most cases require some significant budgetary allocation or expense. Believing and knowing that – as a part of a vast urban network of competing complexities – these special contributions will return as a wise contribution to the overall success of such a creation and a city's heritage.

In the spirit, then, of providing those of us who may be compelled to search further within the realm of urban design – creating a greatness of human endeavor and attaining the success of the human place – we offer a brief contemplation and embracing statement from perhaps our greatest American architect: Louis I. Kahn. His offering was made to us in 1971, at the occasion of his gold medal award acceptance address to the American Institute of Architects:

> How inspiring would be the time when the sense of human agreement is felt as the force which brings new images. Such images reflecting inspirations and put into being by inspired technology. Basing our challenges on present-day programming and existing technologies can only bring new facets of old work.
>
> The city from a simple settlement became the place of the assembled institutions. The settlement was the first institution. The talents found their places. The carpenter directed building. The thoughtful man became the teacher, the strong one the leader.
>
> When one thinks of simple beginnings which inspired our present institutions, it is evident that some drastic changes must be made which will inspire the recreation of the meaning, *city*, as primarily an assembly of those places vested with the care to uphold the sense of a way of life.
>
> Human agreement has always been and will always be. It does not belong to measurable qualities and is, therefore, eternal. The opportunities which present its nature depend on circumstances and on events from which human nature realizes itself.

A city is measured by the character of its institutions. The street is one of its first institutions. Today, these institutions are on trial. I believe it is so because they have lost the inspirations of their beginning. The institutions of learning must stem from the undeniable feeling in all of us of a desire to learn. I have often thought that this feeling came from the way we were made, that nature records in everything it makes how it was made. This record is also in man and it is this within us that urge us to seek its story involving the laws of the universe, the source of all material and means, and the psyche which is the source of all expression.[26]

Case study: Kohn Pedersen Fox – contemporary global urban design project

It is certainly the case that urban design has been significantly transformed by the emerging technologies of the last 150 years. The enabling of multi-story construction with the advent of the elevator and structural steel has been major. These technologies have brought the possibility of adding many significant new vertical functional architectural elements of use, density and form to the urban place.

The noted American architect/urban designer William Pedersen has certainly been one of the major global design leaders in this field. He is the inventive creator of many of the premier international architectural and urban design landmarks of the twentieth and twenty-first centuries. Building on his country's leadership and legacy in high-rise design, when asked, Pedersen well expressed this legacy of spirit and inspiration thusly:

> Louis Sullivan once called the skyscraper a "proud and soaring thing". That was at a time the skyscraper was an anomaly within the fabric of a city. Now, however, the skyscraper is the rule ... not the exception. A "proud and soaring thing" makes a poor urban conversationalist. My tall buildings in Hudson Yards represent the latest development, in over forty years of exploration, of trying to bring this building type into a more social awareness ... where it gestures in response to the participants within its context.

His current design of the Hudson Yards in New York City is the largest urban design project in history and imbues this creative spirit. As being brought about by Pedersen, this project (see illustration) exemplifies many of these new transformations of urban form being brought on through these ever expanding technologies.

Case study: design for community crime prevention – defensible space revisited

The 2015 HBO miniseries *Show Me A Hero* reopened discussion about urban crime, race and the physical environment, viewing the historic Yonkers desegregation case from today's perspective. The HBO series, directed by David Simon, told the story of Yonkers in the 1980s, torn by a federal court decision mandating affordable housing in all parts of the city, in a manner illuminating today's racial equity and police issues. The late Oscar Newman, the noted author of *Defensible Space*, was a key player and favorably portrayed in the movie by Peter Riegert. Newman's concepts regarding the physical design of housing, the nature of public and private space and how it could influence criminal behavior was on display in the effort to locate and build 200 townhouses in different Yonkers neighborhoods. The HBO movie ended with a note that "Newman's ideas have been widely accepted."[27] As discussed in Chapter 7, Community and Social Equity, HUD policy and even the New York City current administration's deep commitment to affordable housing does not support high-rise public housing for families.

Urban redevelopment must often deal with the concerns of crime, both real and perceived. Preventing crime clearly involves factors far beyond what a redevelopment project can control such as: policing, the criminal justice system, and economic and educational opportunity. Yet there is a relationship between redevelopment success and the perception of personal safety. Everyone prefers to live, work and play in places where they feel more secure. Place-making, discussed throughout this book and specifically in G.B. Arrington's Chapter 4, is a key concept of today's urban redevelopment. The type of retail, restaurant and street activities emphasized by place-making hearkens back to both Jane Jacobs' "eyes on the street" and Newman's *Defensible Space* – how activity and community help control behavior, including crime. Newman especially focused on residential neighborhoods, including a 1965 study of the Oaklawn neighborhood in Chicago that was a precursor of efforts to include improving security in revitalization.[28] The following illustration shows how he emphasized the differences and the way residents control private, semi-private, semi-public and public spaces.

While there is no physical design that will completely prevent crime, there are ways in which urban redevelopment projects can be planned that will help improve safety. Technology offers some of the most direct tools: good lighting, surveillance cameras, and communications – such as strong mobile phone service – have an influence. However, what is called target hardening: chains, wire fences and the like are of marginal benefit. Just as good street design discourages speeding (sometimes referred to as traffic calming), good street design can promote activities, sight lines and a perception of safety that becomes a reality. Jane Jacobs and William H. Whyte both wrote about people and activities on a street or in a park discouraging crime. Business people in their street level shops not only watch, but they know their community and will report suspicious activity.

Since Newman wrote *Defensible Space*, there have been many changes in what is now called environmental crime deterrence. That study was among the first to use cameras in vestibules, now cameras are ubiquitous. Wire telephone boxes are no longer needed with

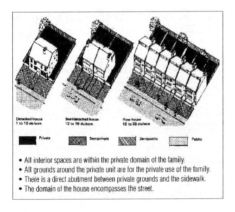

Figure 3.3 Defensible space concept
Source: Oscar Newman.

the advent of mobile cellular phones. Terrorists willing to blow themselves up to damage others are not much deterred by street activity. Rather, they are looking for maximized targeting, and so we now see bollards and other types of hardening to prevent truck bombs. Attacks on theaters and sidewalk cafes target the types of urban places that make attractive communities. Today's redeveloper has to incorporate security into project planning, through design, communication, lighting and policing to protect both life and property within their projects.

Case study: Rocket Street, Little Rock, Arkansas

In downtown Little Rock, Arkansas, the Pettaway neighborhood was a vibrant community in the early twentieth century, but it has subsequently experienced many years of decline. Recently, with civic action and a well-designed project, the neighborhood is improving. The City of Little Rock and the Downtown Little Rock Community Development Corporation have brought the revitalization to fruition, providing affordable housing in an urban area that would serve as a catalyst to the transformation of the entire area.

The neighborhood is bound by Interstate 630 on the north, Interstate 30 on the east, Roosevelt Road on the south and Broadway on the west. This community has been awaiting revitalization for more than a decade; since being devastated by a storm and becoming an area where over 26 percent of the residents live in poverty, almost double the city as a whole (14 percent). In order to spur development in the area, a project calls for a pocket neighborhood, a desirable alternative to urban sprawl. Pocket neighborhoods are clustered groups of neighboring houses or apartments gathered around a shared open space – a garden courtyard, a pedestrian street, a series of joined backyards or a reclaimed alley – all of which have a clear sense of territory and shared stewardship (pocket-neighborhoods.net).[29]

The Pettaway pocket neighborhood project features a creative design that has been recognized by multiple awards – including a 2013 award from the American Institute of Architects. Designed by the University of Arkansas Community Design Center in Fayetteville, the project plan will invigorate the neighborhood with modern carefully designed homes. The property consists of a 1-acre parcel along the east side of Rock Street, between 17th and 19th streets that will accommodate new homes built around a common area and playground, and is located about a mile from the heart of downtown. The design plan calls for differently styled single-family houses: two-story models (1,100–1,400 square feet) and three-story models (1,000–1,250 square feet). These houses will surround a playground and a common area functioning as a shared lawn linking private and public space.

The project planning was funded by 2011 Community Development Grant Block funds from the City of Little Rock and made available by the Downtown Little Rock Community Development Corporation (DLRCDC) with aid by the City as well as the National Endowment of the Arts. Preliminary 2015 cost estimates were put at approximately $1.25 million. Construction funding has experienced difficulties being secured.

The project concept aims to provide housing by incrementally increasing density while avoiding high-rise, multifamily development. It is envisioned as a project with multiple houses that applies advanced measures for storm water control and aims to reduce the carbon footprint of its residents by sharing community lawns, playgrounds, gardens, streets and amenities. A low impact ecological storm water management system is planned, based on landscaping and fixtures designed to take advantage of natural rainfall filtration and avoid polluted runoff into the sewer system.

This type of development, although very attractive in its provisions to take advantage of shared resources and amenities, is not for everyone. There is a compromise between a closed, shared living providing low environmental impact and security with a concomitant decrease in privacy. In order to take this into account, the University of Arkansas Community Design Center team worked with a citizen advisory committee to work towards a desirable product that blends traditional architectural elements with modern principles offering a fresh look – in harmony with the character of the neighborhood.

Case study: Vancouver, British Columbia

Vancouver is renowned for its natural beauty and for urban design. Located between the sea and the mountains, the city is beautiful – but there is limited land. Consequently, Vancouver has adopted strong design controls, which are sometimes referred to as "Vancouverisms." The city has attracted growth and investment from Asia as well as North America, and it has a reputation as a safe and strong real estate market. To manage but allow growth, Vancouver has carefully thought about the location and spacing of high-rise towers. Compare this to New York's "accidental skyline," where changes in technology and zoning allow for the transfer of development rights. This has resulted in a 1,000-foot-tall ultra-luxury condominium development that shades Central Park. Vancouver's urban redevelopment has preserved viewsheds, waterfront access and public open spaces by carefully managing the location of new high-rise projects. The following is a series of projects, all near downtown, that continue the carefully planned growth and urban design approach of Vancouver.

Vancouver, led by retired Planning Director Larry Beasley, has also long encouraged redevelopment of its underutilized industrial areas. The Granville Island redevelopment, started in 1972, was among the first industrial areas to be revitalized with a gritty mix of commercial, artistic and culinary experiences, while viable industries remained. Granville Island remains a highly successful model, while Coal Harbor and False Creek demonstrate two more recent redevelopments in former industrial areas near the downtown.

Woodward's is a successful mixed-use redevelopment located close to downtown, in a neighborhood called Gastown. This project includes the adaptive reuse of the historic former Woodward's department store, education space for Simon Fraser University, two new residential towers, retail and many amenities – all designed by local architects Henriquez Partners. The project was a public-private partnership with the City granting the 2.32 acre site to developer Westbank Projects Corporation. Over one million square feet have been constructed, including 200 units of social housing out of a total of 746 units. Main office tenants include several city and federal government related institutions, as well as Simon Fraser University.

False Creek began in the nineteenth and early twentieth centuries as an important industrial center near downtown Vancouver. The creek itself is short but was teeming with wildlife. The area is framed by highways, and the expansion of downtown and the initial planning was a part of Expo '86. The current proposal is located on one of Concord Pacific's last remaining parcels of undeveloped Expo '86 land and will include eight residential buildings with over 1,000 units for 3,000 people. The area will also include 90,000 square feet of amenities, bringing a selection of amenities and commercial activity such as a public health and fitness center, restaurants and lounges, cafes, a market, boutiques and more. Concord Pacific's plan also includes a new bicycle and pedestrian route that will connect downtown Vancouver to the seawall and the Cambie Bridge. Completion of all the towers, amenities and infrastructure is not expected until 2023.

Figure 3.4 Vancouver skyline
Source: Freeimages.

That neighborhood will see some major changes in the coming years. The BC Place casino project was re-announced recently as a $535 million entertainment and urban resort destination. However, a smaller casino is planned rather than the one previously proposed to simply replace Edgewater Casino. While casinos have been a part of quite a few redevelopments (see Chapter 2 discussion of historic Bethlehem, PA, steel mill), there are those in Vancouver and elsewhere that worry about the overall impact of gaming, both economically and socially, on a community.

Major residential and commercial developments are also planned for the site of the Plaza of Nations – a parcel of Concord Pacific-owned land to the west of the planned expansion of Creekside Park. Three rental towers have been approved for the Rogers Arena and will bring 614 rental units to the downtown core. The first tower is currently under construction and its lower levels will include a major expansion of the Rogers Arena concourse – plus additional food, restaurant and bar options. The project has been touted as the largest rental-only development the city has seen in nearly thirty years.

Another project, The Rise, has been called an exceptional model of transit-oriented, central city development that successfully mixes large-format retail uses with smaller shops and housing in a well-resolved mid-rise form. Grosvenor Americas' mixed-use, residential-over-retail building is located just south of downtown Vancouver on Cambie Street, a major north–south arterial road and transit route. The building occupies an entire 2.3-acre block and includes 92 rental live/work units above 200,000 square feet of retail space, a 1-acre fully internalized truck court and waste/recycling area, and 3 levels of underground parking with 520 retail and 121 residential parking stalls. The retail component includes three large-format stores, each situated on its own floor. Smaller street front shops line two sides of the building. The residential component sits on the roof of the retail podium in an open-air townhouse and flat configuration – with units surrounding a functional, shared, grass-covered courtyard and community garden.[30]

A two-year study is underway to plan the removal of the Dunsmuir and Georgia Viaducts. With growth, new developments and the diversion of traffic volumes into Pacific Boulevard, the removal of the viaducts has many worried of severe impending traffic congestion. There are a number of alternative transportation options being created in Vancouver, such as the noted bikeway design for the Grey-Cornwall project.[31]

The new Vancouver Convention Center West nearby in Coal Harbor, designed by LMN, Architecture, Urban Design and Interface with MCM+DA, is another new redevelopment project that is based upon ecology-based land use featuring the 1,000-acre Stanley Park. Features include the largest (6 acre) green roof in Canada, including 400,000 indigenous plants and 240,000 bees, with a sloping form that connects to Stanley Park, a view across the Burrard Inlet. Functionally, the large permeable surface controls rain-water and slows storm water runoff, and helps reduce the heat island effect in the downtown core. Adjacent property owners had a vested interest in the roof's appearance and contributed to workshops on the building's "fifth elevation" visible from nearby high-rises. City zoning ordinances strongly shaped the forms, requiring that the building maintain view corridors from downtown streets out to the water. The final form of the roof modulates to extend the lines of downtown streets and adds a new view corridor towards the Northwest.

Notes

1 Porter, Paul R. and David C. Sweet, *Rebuilding American Cities*, Center for Urban Policy Research, 1984.
2 Boston Redevelopment Agency.
3 Burchell, Robert, Center for Urban Policy Research, Rutgers University, 2008.
4 Barnett. Jonathan, *City Design*, Routledge, 2011.
5 Glaeser, Edward, *The Triumph of the City, How Our Greatest Invention Makes Us Richer, Smarter, Greener, Healthier and Happier*, Macmillan, 2011 and Chakrarbti, Veershan, *A Country of Cities*, Metropolis Books, 2014.
6 McLennan, Jason F., *The Philosophy of Sustainable Design: The Future of Architecture*, Ecotone, 2004.
7 Bruntland Commission, formerly the World Commission on Environment and Development, 1987.
8 Bacon, Edmund, *Design of Cities*, Viking Press, 1967.
9 www.miami21.org/typesofzoningcodes.asp.
10 Duany, Andres, *The Smart Growth Manual*, McGraw-Hill, 2010.
11 Americans with Disabilities Act of 1990 – ADA – 42 U.S. Code Chapter 126, 1990 as amended.
12 Gruen, Victor, *The Heart of Our Cities*, Simon and Schuster, 1964.
13 Hecksher, August, *Alive in the City, Memoirs of an Ex Commissioner*, Charles Scribners Sons, 1974.
14 Johnson-Marshall, Percy, *Rebuilding Cities*, 1966.
15 Park, Robert and Ernest W. Burgess, *Roderick Duncan McKenzie*, University of Chicago Press, 1925.
16 Hoyt, Homer, *The Structure and Growth of Residential Neighborhoods in American Cities*, Washington, Federal Housing Administration, 1939.
17 Lynch, Kevin, *Image of the City*, Harvard-MIT Joint Center for Urban Studies, 1960.
18 Alexander, Christopher, *A City Is Not a Tree*, Architectural Forum, April/May, 1965 and Alexander, Christopher, Sara Ishikawa, Shlomo Angel and Murray Silverstein, *A Pattern Language: Town, Buildings*, Construction, Oxford University Press, 1977.
19 Op. cit. Glaeser, Edward, *Triumph of the City: How Our Greatest Invention Makes Us Richer, Smarter, Greener, Healthier, and Happier*, Macmillan, 2011.
20 Rudofsky, Bernard, *Streets for People*, Doubleday & Company, 1969.
21 Mayer, Harold M. and Richard C. Wade, *Chicago: Growth of a Metropolis*, University of Chicago Press, 1969.
22 Op. cit. ADA, Americans with Disabilities Act of 1990 – ADA – 42 U.S. Code Chapter 126, 1990 as amended.
23 Safdie, Moshe, *Moshe Safdie II*, Architizer, 2009.
24 Zucker, Paul, *Town and Square, From Agora to the Village Green*, Columbia University Press, 1966.
25 Blumenfeld, Hans, *The Modern Metropolis*, MIT Press, 1971.

26 Kahn, Louis, *Writings, Lectures, Interviews*, Rizzoli, 1991.
27 *Show Me a Hero*, The HBO Miniseries 2016, based on the book by Lisa Belkin, 1999.
28 Newman, Oscar, *Defensible Space*, The Macmillan Company, 1972 and also Parkmall Lawndale, Washington University, 1966 and cited: academia.edu/11309615/The_Economy_of_Fear_Oscar_Newman_Launches_Crime_Prevention_through_Urban_Design_1969-1970.
29 Pettaway Rocket Street Neighborhood Design, AIA Award 2012 http://uacdc.uark.edu/work/pettaway-pocket-neighborhood.
30 Urban Land Institute Case Study, May, 2014.
31 Next City, Bikeways, December, 2015.

4 Transportation

G.B. Arrington

People moving to city shaping

Transportation is not just about moving people from one place to another. In urban redevelopment it's also about leveraging our transportation system to create places where we want to go. It's about transforming what we already have to get what we actually want. Creating the kind of transportation and places, especially urban redevelopments that add value and livability to our communities; focusing less on the measures of travel like efficiency and congestion, and more on outcomes such as helping to create and enhance our communities.

Transformative transportation focuses on the outcomes we desire not just the trip. Harnessing transportation to help shape the kinds of places we want to live, work, play, shop and learn. In some ways this is an old story – in the 1920s, Henry Huntington's interurban Pacific Electric lines and land development companies intentionally reshaped the landscape of Los Angeles. Naples, Seal Beach, Huntington Beach, Newport Beach and Redondo Beach stand in testament to the transformative power of transit investments to shape cities and move people.[1]

The Portland Oregon region and the Rosslyn Ballston Corridor in Arlington County Virginia are contemporary examples of intentionally linking major rail transit investments

Figure 4.1 SF.Embark.MuniXXX Embarcadero & Brannon Street Station, San Francisco, CA. Leveraging transportation in urban redevelopment to create the kind of places we want to be, such as this station and new housing overlooking San Francisco Bay where an elevated freeway once stood, requires applying new measures of success

Source: Photograph by G.B. Arrington.

and a vision for how to grow to shape cities and move people. They stand out as examples in part because they are the exception not the rule for transportation. Portland and Arlington County excelled by being willing to think about transportation differently. To harness transportation investments to help create the places they desired and thereby enhance the quality of life and vitality of their communities.

The art and science of harnessing transportation to support the redevelopment and enrichment of our cities starts with reconceiving how we think about transportation. What we expect to see from our transportation investments. What we solve for and how we measure success. It has been said that before. It has been said that before Christopher Columbus could discover the New World, he had to unlearn that the world was flat. Like Christopher Columbus Portland and Arlington County needed to unlearn much of the traditional wisdom they already knew about transportation and urban redevelopment before they could chart a new transformative path.

Part of the relearning process in Portland and Arlington was coming to understand that solving for a single measure like congestion would not yield the outcomes they desired. That addressing congestion was treating the symptom not the root cause, something the medical profession calls symptomatic treatment. That simply adding lanes was the medical equivalent of the old phrase "take two aspirin and call me in the morning." Unfortunately much of our conventional knowledge about transportation has led us down a road where the preferred two aspirin "solution" ends up creating even bigger problems. This is not just a challenge for how we've built our highways and roadways. It is a systemic problem within the transportation industry – add two lanes and call me in the morning.

Modern streetcars

Modern streetcars are a great example of rethinking transportation. Of using transportation to address a root cause of the aliments driving many urban redevelopment projects – rethinking transportation to achieve desired land use and economic development outcomes. Depending on where you sit, what you value and what you are solving for they are a wonderful idea or perhaps a misguided one. As central city redevelopment projects streetcars have excelled in cities such as Portland and Seattle, Washington where they have been

Figure 4.2 South Waterfront, Portland, OR. Modern streetcars have been used by cities such as Portland and Seattle as a catalyst to help drive the redevelopment of former industrial districts in close proximity to their downtowns

Source: Photograph by G.B. Arrington.

catalysts for billions of dollars in transit-supportive real estate development as part of a broader city shaping strategy.

The Portland Streetcar was the first modern streetcar system built in the United States when it opened in 2001. The Portland Streetcar is a case study of leveraging local, state, and federal resources to link transportation investments and real estate development. In Portland the streetcar has been credited with leveraging more than $5 billion in new public and private development within two blocks of its route.[2]

The Seattle streetcar project opened in 2007 and travels from the downtown retail core through the Denny Triangle and South Lake Union (SLU) neighborhoods to a waterfront park on Lake Union. The scale and type of development that has occurred in the SLU over the last twenty years is striking. The area has experienced $3.2 billion in new development in the neighborhood with mixed-use since 2007.[3] Perhaps most striking is Amazon's new headquarters. The $1.16 billion headquarters will include a trio of thirty-eight story towers as part of eleven new buildings. Amazon is providing $5.5 million to purchase a fourth streetcar, subsidize more frequent service for ten-years and bicycling improvements to compensate for taking city-owned alleys for its high-rise campus in Seattle's Denny Triangle.[4]

In terms of speeding traffic and people movement these light rail systems have been criticized as performing no better than buses running in mixed traffic. So are they transportation projects or are they redevelopment projects or are they a bit of both?

What is really illuminating about the streetcars example comes with peeling back the onion and seeing who is asking the question about what is important; what are really the problems we are trying to address with our transportation projects. It used to be that "transportation people" alone asked the question. But a funny thing happened on the way to the twenty-first century. Increasingly cities have been asking the question about what is important with our transportation projects. When different groups ask the same question the ultimate answer is likely to be nuanced; more complex, more a balancing a number of measures.

Just like a single roller-coaster on a hill does not make a Disneyland, a modern streetcar alone does not make a successful downtown. But a streetcar can be a powerful city shaping tool when linked with a vision for the future, an expanding real estate market, strong leadership and supportive planning and zoning. Harnessing the city shaping capacity of streetcars only became possible when cities came to the transportation table with a different set of questions and desires than the single measures being evaluated by the usual suspects in the transportation community.

Two different paths to a twenty-first-century metamorphosis

Just as Arlington and Portland set out on a different path as twentieth-century pioneers by rethinking transportation, the Bay Area Rapid Transit District (BART) and Tysons Corner Virginia provide twenty-first-century works in progress of rethinking transportation. While on one hand they are very different – one is a regional transit system and the other a suburban edge city – on the other hand they share a common thread. Both grew to success out of transportation and both came to realize that to enjoy future success they needed to reinvent themselves with different land use and different transportation.

Figure 4.3 Fruitvale Village viewed from the BART Platform, Oakland, CA. Sometimes the path to success with urban redevelopment means reinventing yourself with different land use and different transportation. At Fruitvale, BART's surface commuter parking became a mixed-use transit village

Source: Photograph by G.B. Arrington.

BART's journey into the twenty-first century

To succeed in the twenty-first century, BART is in the process of reimaging, reinventing and rebuilding itself. With its roots in the 1960s BART was planned to serve the bygone needs of a different era. When viewed through today's contemporary glasses BART is a bit of an enigma – an auto-dependent transit system in a region heavily invested in focusing walkable communities around transit. With 40 percent of all household growth targeted to be near a BART station[5] the regions adopted growth strategy is betting heavily on BART.

Accommodating 166,000 new households near its stations will require BART to be a different kind of a neighbor. BART will need to wean itself from being an auto-dependent transit system to one that enables, facilitates and embraces new transit-oriented communities in and around its stations. That will be a tall order. With more than 46,000 parking spaces it should be of no surprise 49 percent of BART riders get in a car and drive to BART from their home. BART will also need to become the mode of choice for all of the communities it serves. Of the 400,000+ trips made at BART's forty-five stations each weekday, more than 60 percent of those trips originate or end at three stations in downtown San Francisco.[6]

BART was conceived and designed to be a traditional commuter rail system. Now the expectations for and the role of transit are very different in the Bay Area. As BART reinvests in its capital facilities, focusing on moving commuters into San Francisco is no longer sufficient. Meeting those future needs comes with a very big price tag. In 2014 BART estimated it faces nearly $20 billion in operating and capital needs over the next ten years for system reinvestment and expansion.[7]

Making BART stations good neighbors for the 160,000 households targeted around its stations in Plan Bay Area means changing parts of BART's design template and paying more attention to what happens outside its fare gates. That means balancing BART's people moving function and its city shaping function. Bringing those objectives to scale at BART will be shaped by three building blocks:

1 **Make Transit Work** – Consider the role the station will play in the community and how tweaking transit access design (buses, cars, bikes and pedestrians) can make transit work better for tomorrow's requirements and be instrumental in positively shaping a community's future. Sometimes that may require breaking the mold of generally accepted transit design to address transit function and the community's vision.
2 **Create a Place** – The area immediately around a station can become destinations with multiple functions. It can support transit functions, new development, provide open space, public art, new community services or be an architectural landmark providing new experiences.
3 **Connect to the Community** – The best fitting clothes are tailored, designed to fit each unique shape. Similarly, the best transit stations are tailored to each unique community in which they exist. For BART that means stepping outside their property lines and blurring the edges of their facilities to better link to, complement and be a good neighbor to the communities they serve.

The BART example stands out because its leaders came to understand its past was not necessarily its future. That to continue to be relevant BART needed to redevelop its physical plant with an eye to a future. BART needed to evolve from an auto-oriented transit system to more of a community-oriented system that serves as a catalyst for city building and a vehicle for people moving. By taking those steps BART will be better positioned for the future by meeting the changing needs of its riders, employees, the community and its partners.

Tysons Corner: from Edge City to twenty-first-century city

Tysons Corner Virginia grew-up around the automobile; it is now being transformed and transported into the future around its four new Metrorail stations which opened in 2014. The Tysons Task Force report *Transforming Tysons*[8] charted the course for the redevelopment of the nation's twelfth largest employment center into a new downtown for Northern Virginia. The 2008 plan reflects the hopes and dreams of the businesses, residents, neighbors and stakeholders invested in Tysons future. It provides for a place that will not be simply bigger, but better.

The fundamental redevelopment of Tysons is now well underway – from a national poster child for suburban sprawl and automobile dependence into world-class downtown exemplifying principles of TOD and green neighborhood design. Tysons' 1,700 acres are evolving from forty-six million square feet of development and forty million square feet of parking into a twenty-first-century city of 160 million square feet of livable, walkable, mixed-use, transit connected, green urbanism.

A major pivot point in the reinvention of Tysons came through demonstrating that by growing in a different way Tysons could grow larger and have much less congestion than the path it was on. By reinventing land use and transportation, the initial analysis indicated that increasing development by 160 percent would only cause a congestion increase of approximately 11 percent.

Four key drivers underpin the Tysons transformation strategy:

1 **First, substantially increasing the housing in Tysons** to get a better housing/jobs balance – from 17,000 residents to 100,000 residents – completely through urban

redevelopment. The twenty-five-floor Ascent at Spring Hill Station completed in 2014 is the guinea pig for new residential high-rises in Tyson. A number of residential towers are underway in its footsteps.

2 **Second, focusing growth around Tysons' four metro rail stations** and a network of internal circulators. The adopted comprehensive plan, new zoning and urban design guidelines reflects the plan, which located 95 percent of all growth within a three-minute walk of four new Metrorail stops, and three planned circulator transit routes.

3 **Third, creating a tight grid of interconnected streets** favoring walking, biking, transit and local auto-trips. A grid of streets transportation funding plan supported in part by an assessment district was established in 2014 and the first grid street has been completed.

4 **Fourth, greening Tysons with a multifunctional green network** will be created where none exists today. With over half the land area dedicated to streets and parking currently, 160 acres of new parks and the restoration of two streams will provide balance to the increase in development. Fairfax County has secured a number of public facilities, parks, and active recreation fields in Tysons, as recommended in the Plan through the use of development proffers.

Large-scale urban redevelopment like Tysons comes with high infrastructure costs and impacts. The balance point in Tysons was finding the right urban scale and mix of land use, creating a more connected grid of streets and including a package of new amenities to get a consensus from the neighbors, landowners and developers, and the environmentalists. That directly translated in making it possible for Tysons to support more growth without more impacts. The neighborhoods were concerned about the spillover impacts of more growth in Tysons on their quality of life. For landowners/developers the balance was having the plan include enough new growth to make it financially feasible to foot much of the bill to pay for the transportation and amenities. The smart growth community were solving for environmental stewardship and creating dense walkable places surrounding Tysons' four new metro stops.

Two assessment districts have been established to help fund Tysons implementation. The first provided a $400 million contribution towards the capital cost of the new metro stations and the second provided $250 million towards a grid of new streets and other improvements essential for rebuilding Tysons into an urban place. The first district was established in 2004 and the second in 2014.[9]

Private sector plans are currently in motion to expand Tysons by 80 percent. Fairfax County has granted approval for nearly 22 million square feet in development projects consistent with the plan and is reviewing an additional 14 million square feet.[10]

Conclusion

Communities are increasingly approaching transportation projects such as major transit investments as part "people-moving" and part "city-shaping." It is in many ways an old story guided by new sensibilities. Taking a "development oriented transit" approach, as the Portland region describes it, requires balancing both aspects throughout the transit facility design, corridor selection and station location decisions to optimize transit operations, community fit, urban design and economic development.

Transportation case study: the Pearl District – Portland's largest TOD

Portland's Pearl District is a shining example of the transformative power of transportation, supportive public policy, sustained public-private partnerships and market demand for walkable urban places. "The Pearl" is a transit-oriented district spanning approximately 100 city blocks bounded by I-405 to the west, West Burnside Street to the south, NW Broadway Street to the east, and the Willamette River to the north (it is north of and adjacent to Portland's CBD). In 2008, 58 percent of residents reported using modes other than driving to get to work.[11] Once a rail yard and an "incubator" for start-up businesses in abandoned warehouses, and home to a large artist community, the Pearl District is now Portland's largest and arguably most successful mixed-use neighborhood. In a series of complementary public and private actions the Pearl was purposefully designed around transit, walkability and a mix of uses. A major catalyst to the transformation of the Pearl District was the construction of the Portland Streetcar, the first modern streetcar system to be built in the United States.

Development oriented transit

In many ways the Pearl District can be explained as a variation on the well-tested Portland strategy of linking transit and land use. Since the 1970s with the construction of the Downtown Transit Mall the region has strategically used major transit investments as a catalyst to leverage its land use and economic development objectives. That pattern has been repeated over nearly four decades with the Transit Mall, a succession of six light rail lines and multiple extensions of the Portland Streetcar.

Part of the essential alchemy of the Pearl District was understanding that contemporary transit is part people moving and part community building – and depending on the situation the design of the transit project needs to reflect the balance point between the two. From the beginning, the Portland Streetcar was always more about community building than people moving. Simply having a streetcar was not enough to complement the lifestyle,

Figure 4.4 Tanner Springs Park, Portland, OR. Tanner Springs Park is emblematic of the nuanced public-private partnership that resulted in Portland's largest and most successful TOD. Land for this and two other parks were dedicated to the city by developers as part of a 50-year development agreement

Source: Photograph by G.B. Arrington.

image and the "pedestrian accelerator" the developers of the Pearl were seeking. To play its part as a real estate catalyst the streetcar needed to be conceptualized, designed, constructed and operated with development very much in mind. That perspective shaped key decisions on the streetcars operations plan, route selection, station design and location to optimize community fit, urban design and economic development.

That emphasis on development over operating speed has led some transit advocates to criticize the streetcar for placing too much emphasis on development and not enough on transit. Despite the criticism the Portland Streetcar excels in attracting riders. With an average daily ridership of 15,055[12] in February 2015, the Portland Streetcar has the highest daily ridership of any new streetcar line in America.

Public and private initiatives shaping the Pearl District

Planning for the area began with the 1972 *Downtown Plan*. The Downtown Plan recognized the important supporting role of the north downtown area as an industrial and distribution center. At the same time, the Plan also acknowledged changing development patterns and recommended replacing some industrial uses with mixed-use development.

The 1988 *Central City Plan* built on the work of the Downtown Plan, extending its geographic scope and expanding its range of policy concerns. It established the Central City Plan District, which includes the Pearl District. The Central City Plan illustrated the intended changes for the industrial area from railyards to a residential/commercial area. To facilitate this transformation, the Plan retained existing industrial zoning but allowed central employment zoning when services could be provided; adopted residential district zoning regulations; and allowed use of FAR (floor area ratio) residential bonus provisions.[13]

Other urban infrastructure and sustainability

While transportation is most tightly tied to urban redevelopment, some projects also benefit greatly from being able to utilize existing urban systems such as water, sanitary sewer, storm water, electricity, and telecommunications. In terms of smart growth, redevelopment as opposed to suburban sprawl, avoids the need for expensive expansion of systems and the loss of farm or open spaces. While in many cases these systems must be upgraded, repaired or expanded to meet modern standards and to carry the increased demands of redevelopment the costs and impacts may be significantly lower.

A common issue, especially in the older cities of the eastern United States and Canada are combined sewer outflows. These systems use the same pipes to carry storm and sewer water to treatment plants, but when it rains the plant is overwhelmed, and the system discharges both sewer and storm water directly into a water body. This is a difficult problem and expanding treatment plant capacity or engineered retention are very expensive and not an ideal solution. Urban redevelopments today often incorporate greener, more sustainable and softer systems that use swales, plantings and landscape treatments such as rain gardens, which effectively moderate storm water flows.

Waterfront redevelopments face an additional set of infrastructure related challenges, starting with resilience to flooding and storm water outflows during a storm. While there is great real estate opportunity to being on the water, there are also risks that must be managed. These projects are subject to review by many more agencies; wetlands and the US Army Corps of Engineers, flood plain ordinances and FEMA (Federal Emergency Management Agency), and ecological concerns involving US Fish and Wildlife. The design of the water's edge is especially crucial, with the current emphasis on softer, more ecological and resilient designs rather than rigid barriers such as sheet-piling or riprap. Not only does the infrastructure need to work, reduce risk, and be cost-efficient, public access to the waterfront is key to success in gaining approvals and in marketing.

The new buildings of urban redevelopment, often meeting so-called green standards (LEED in the US, Green Globe in Canada) have numerous characteristics that improve sustainability and minimize impacts on infrastructure. Building equipped with plumbing systems that reduce sewer flow, from faucets in office buildings that automatically turn off, to low-flow toilets, and low water use washing machines in residential units, all reduce the sanitary sewer demands. Large, sophisticated buildings may use so-called "grey-water" systems that recycle all but toilet water. Green roofs and landscaping, and less impermeable surfaces for parking, all minimize storm water flows.

The same approach applies to the electrical grid, as modern urban redevelopment features high-energy performance buildings. Again, green standards and cost effectiveness require more efficient systems; HVAC equipment and improved building siting and insulation values, LED and other efficient lighting that result in less electrical demand. Alternative energy systems are increasingly used in urban redevelopment. These include solar-voltaic panels to generate electric, solar hot water for residential homes, wind power generation and geothermal systems can all reduce energy demand, and benefit from existing electric infrastructure with one backing up the other.

An important variation on how infrastructure is thought of as part of urban redevelopment today is the issue of resilience and the use of micro-grids. Experience has taught us all that storms, earthquakes, terrorist attacks and electric power failures all have devastating effects upon those who live in urban areas. One form of resilience planning is to have micro-grids, smaller local systems for either power generation or telecommunications that allow a community to function effectively even if the power grid goes down for any reason. Cogeneration, most often today generating electricity and steam from natural gas, can not only reduce power demands especially at peak times, but also serve as a back-up. A base router with its own emergency power supply can keep critical systems such as telecommunications functioning. Red Hook, a waterfront neighborhood in Brooklyn, New York was inundated and without power after Superstorm Sandy but a nascent micro-grid system helped communication.[14] San Diego has supported the Borrego Springs micro-grid of renewable energy, allowing a neighborhood to have both low-cost and resilient electricity.

This type of technology is also part of the trend towards what is called Big Data which many technology experts see as key to so-called Smart Cities. Just as marketing firms know a great deal about prospective purchasers, technology allows city planners and engineers to monitor everything from traffic to sewer flows; where people, vehicles, water, energy are moving and to adjust systems accordingly. Some cities, usually larger, have data centers where traffic, police, water and other real time is portrayed and resources can be allocated where needed.

The provision of updated and expansion infrastructure, from light rail to micro-grids, is a key component of urban redevelopment. Reuse, improvement and expansion of existing systems is often, but not always, less costly than expanding service to sprawling suburbs. The ability to use existing rights of way alone can be a significant advantage. Just as rehabilitating an old building is often more complicated and less routine than new construction, the weaving together of the infrastructure network may take many meetings and agreements than sprawl, the results can be more sustainable, authentic and unique.

Case study: Denver TOD – the next big thing?

The scale of what is unfolding in the Denver Region with transit expansion and urban redevelopment through Transit Oriented Development (TOD) is unprecedented in modern American history. The protesters at Chicago's Grant Park in 1968 might have been talking about Denver while they chanted, "the whole world is watching." With some sixty new transit stations and five transit lines coming online in 2016 through the FasTracks transit expansion program the Denver Region has an opportunity no modern American city has been able to realize – to build a regional rail network and link it with land use planning to accommodate growth without diminishing livability.

The initial signs indicate the Denver Region is heading down the path of siezing their TOD opportunity. Fueled by a strong real estate market and supportive public policy over 31,000 new housing units and twenty-eight million square feet of government and commercial space have been built or are underway along the Denver Regional Transit District (RTD) system since FasTracks passed in November 2004.[15] Denver Union Station (DUS) has shepherded a new transit-oriented downtown neighborhood. The opening of five new rapid transit lines and 50+ new stations in 2016, changing real estate market preferences for urban living, the adoption of local land use plans encouraging TOD at virtually every RTD station and the attractiveness of the Denver Region as a place to live, work and play all point to the inevitability of more and more TOD.

Incrementally, then boldly building a regional rail system

Following the lead of other cities (San Diego, Portland, Sacramento, San Jose and Los Angeles) with new light rail transit (LRT) systems Denver RTD has been following a well-trodden path of incrementally expanding their rail system line-by-line rather than trying to build a new system all at once. RTD opened the 5.3-mile long Central Corridor LRT line in October 1994. The Southwest Corridor line followed in 2000 and the Central Platte Valley

Figure 4.5 Mariposa TOD at 10th & Ossage Station Denver, CO. This mixed income TOD is an example of the uptick in development activity currently underway at RTD's stations. With 800 units and community facilities Mariposa is expected to be built in nine phases by the Denver Housing Authority

Source: Photograph by G.B. Arrington.

spur in 2002. In November 2006 the Southeast Corridor opened. Then by 2013 with the opening of the W line, the first of six new lines funded by FasTracks, Denver's rail network had grown to six lines and forty-six stations.

FasTracks fundamentally changed the pace, scope and reach of transit expansion in the region. FasTracks is the funding measure supported by 58 percent of Denver Region voters in 2004 to expand the sales tax by 4 percent to build what is now the $7.8 billion FasTracks plan – six new corridors, 122 miles of rail, eighteen miles of Bus Rapid Transit, new park-n-rides, expanded bus service and the redevelopment of DUS. The yes campaign had the support from all forty-one mayors plus the Chamber, industry and environmental groups.

Seven years earlier voters rejected a conceptual transit ballot measure opposed by RTD's dysfunctional directly Board of Directors with a 58 percent no vote. The failed Guide the Ride vote in 1997 prompted backers of a metro-wide transit system to redouble their efforts. In 2001–2003 a comprehensive, region-wide FasTracks plan was developed in a highly inclusive process. The resulting map with all the improvements was a centerpiece in the campaign. The campaign stressed, "The time is now" you "Can't stop growth – need to plan for it." Notably one of two daily papers, the Governor and Executive Director of the Colorado Department of Transportation opposed the campaign.[16]

TOD evolution: from city with transit to transit city

To put Denver's experience in context it's useful to quickly step back. Like many of its national peers, the Denver Region is in the midst of an evolutionary multi-decade long city shaping journey as it collectively prepares to leverage the opportunity afforded by the FasTracks investment. Where Denver will end up remains an open question, but the early prognosis is encouraging.

At the end of the day the true test for a region such as Denver is not how much development occurs next to transit. Rather the test is the extent to which the development next to transit is shaped by transit in terms of its density of development, mix of uses, design for walkability and whether the amount of parking provided reflects the presence of transit. To put it another way, you might say for illustrative purposes within the family of TOD there are two "brothers" – TAD and TOD. Transit Adjacent Development (TAD) can be described as development in close proximity to transit. TAD comprises the majority of development near transit in the US.[17] TOD on the other hand is development that is "oriented" to and shaped by transit. The challenge for Denver will be whether they can deliver TODs to their station areas and not TADs.

There has now been enough experience with creating those transit shaped places through ongoing planning and implementation nationally in places such as Washington, DC and Portland to describe what Denver is going through as a four phased process of evolving from a city with transit to a transit city. In other words, to becoming places where the walkable areas around transit stops are transformed into highly desirable sought after districts shaped by and oriented to their transit station. The planning and implementation journey from a city with transit to a transit city moves through these four phases:

1. Getting started
2. Getting plans and policies in place
3. Focusing on implementation
4. Transforming your region

For Washington and Portland that journey towards transformation has consumed much of four decades. One of the first steps in Washington, DC was the creation of a "Transit Development Team" in October 1970.[18] Similarly Portland started in the late 1970s. As a result before construction started on Portland's first LRT line in the early 1980s every station area had been rezoned to encourage transit-supportive development. What Portland and Washington planners did not understand at the time was that planning for TOD was necessary but not sufficient. To achieve meaningful TOD investments their regions would need to go beyond simply planning for TOD and focus on implementation by building new tools and bringing new partners to the table.

Experience reveals guiding, shaping and developing TOD is complicated, typically no one entity has all the tools required to enable TOD. Transit agencies, such as Denver RTD, typically play an important early role in educating and advocating for TOD. Over the years RTD's role as a TOD advocate has evolved considerably as has its partners. RTD went from a passive transit agency TOD program with one staff person when its TOD program was created in 2000 emphasizing "the T in TOD" to a staff of three directly involved in development projects guided by its 2010 TOD Strategic Plan and propelled by the reach of FasTracks. RTD's local government partners have followed a similar evolutionary path.

Early TOD planning in the Denver Region

Since the Denver Region came relatively late to the LRT party it had the luxury of being able to learn from and apply the experience of early adaptors. As the politics surrounding building a regional rail system coalesced, so did the pace of planning for TOD. With the 2004 passage of FasTracks TOD planning also accelerated. Between 2006 and 2010, in what may be the largest undertaking of its type in the country, local governments at existing and future RTD stations undertook forty Transit Station Area Plans across the region. Examples of early TOD planning activities in the region include:

Table 4.1 Four phases of TOD evolution

Phase	Typical activities
TOD 1.0 Getting started	Education and advocacy Collation building Adopt supportive policies Best practices library
TOD 2.0 Getting plans in place	Plans and policies adopted TOD is now legal Focus on residential TOD and a few select sites Transit agency lead
TOD 3.0 Implementation focused	More district based TODs Many plans require TOD reshape transit for TOD Robust TOD tools City leadership
TOD 4.0 Transforming your region	Market prefers TOD Green, equity and jobs TOD Tight parking ratios TOD part of regional DNA

- **City Center Englewood 2000** – the Regions first TOD opened replacing Cinderella City a dead regional mall. The City of Englewood purchased the 55-acre site along the SW LRT line, selected a master developer and rezoned the site; the project includes a new City Hall, 438 residential units and a Wal-Mart furthest from the station.
- **Blueprint Denver 2002** – a citywide integrated Land Use and Transportation Plan adopted as part of Denver's Comprehensive Plan promoting mixed-use development and higher density at stations. The three basic themes within the plan were: (1) areas of change and areas of stability, (2) multi-modal streets and, (3) mixed-use development along main streets, TOD around rapid transit stations, town centers, and other urban centers.[19]
- **Denver TOD Strategic Plan 2006** – a guide for prioritizing the planning and implementation activities of the City and County of Denver related to transit planning and transit-oriented development. The plan created a TOD Typology to help distinguish the types of places that will be linked by the transit system and frame expectations about the mix and intensity of development at specific stations.[20]
- **City of Denver TOD initiative 2006** – following the Strategic Plan, the City hired two teams of Portland based consultants to complete TOD plans at eight stations. The plans defined the overall vision, specific land use mix, circulation patterns, urban form, open space and other public amenities for the area within walking distance of a station. The plans were adopted and the stations rezoned in 2010.[21]
- **Arvada Station Framework Plans 2007** – In 2005 the City of Arvada started the process of undertaking TOD plans for their three FasTracks station areas. The plans followed the same general framework as the TOD plans completed in Denver.[22] The City followed up the Station Framework Plan with an Urban Renewal Plan, Pedestrian and Bicycle Plan, Parking and Transportation Demand Plan, and Zoning District Guidelines. In 2010 Arvada partnered with RTD on a TOD Pilot to help implement their plan for the Old Town Station.

Figure 4.6 Built on the site of the former Alameda Station Park-N-Ride this 275 unit market rate multifamily apartment and townhome project is separated from the station by a plaza. The project was one of the TOD pilots undertaken by RTD

Source: Photograph by G.B. Arrington.

New tools, new partners and new goal posts

Following the initial planning forays involved in TOD 1.0 and 2.0 the Denver Region has moved into TOD 3.0 – a heightened emphasis on implementation – by making an important course correction, developing new tools focused on implementation and welcoming new partners to the table. Where in its earlier foray Denver could largely borrow from other regions, to move forward, the region made a concerted effort that involved some selective borrowing and some bold new homegrown innovations.

Denver's transition into implementation also coincided with a shift within part of the planning world on what success for TOD looked like. Led by a coalition of philanthropic and non-profits focusing on Denver, the Twin Cities, Atlanta and the San Francisco Bay Area the case was being made and tools were being developed to make TOD equitable. They pointed out TOD had emerged as a powerful tool for creating livable communities. And as demand for livable communities grew, land values near transit increased, which could sometimes lead to gentrification and more specifically displacement of existing residents. It was no longer enough to build TOD, regions like Denver needed to also pay attention to making it equitable.

The coalition explained it this way:

> Equitable TOD prioritizes social equity as a key component of TOD implementation. It aims to ensure that all people along a transit corridor, including those who are low income, have the opportunity to reap the benefits of easy access to employment opportunities offering living wages, health clinics, fresh food markets, human services, schools and childcare centers. By developing or preserving affordable housing and encouraging locating jobs near transit, equitable TOD can minimize the burden of housing and transportation costs for low-income residents and generate healthier residents, vibrant neighborhoods and strong regional economies.[23]

The tools the Denver Region has been creating in its TOD 3.0 phase balance the types of TOD implementation tools created in other regions, plus the homegrown invention and creation of a new set of tools to address equitable TOD. The scope and breadth of the initiatives to specifically advance TOD spanning local governments, the transit agency, and the philanthropic and non-profit communities is particularly illuminating:

- **Denver TOD Fund 2015**. In 2010 The Urban Land Conservancy (ULC), Enterprise Community Partners, the City and County of Denver, and several other investors partnered to establish the first affordable housing TOD acquisition fund in the country. The fund was initially capitalized at $13.5 million to purchase and hold sites for up to five years, in anticipation of new transit stations opening with FasTracks. The fund was recently expanded to $24 million and its geography expanded to cover sites within ½ mile of rail and ¼ mile of frequent bus within the Denver Region.[24] According to one of the fund partners "the Fund answers a basic real estate conundrum: when the economy is bad, property values are low and ripe for purchase, but access to capital is poor and affordable housing developers are scarce."[25]
- **ULC**. The Non-Profit ULC was described by *The Atlantic* in 2014 as part land bank, part community focused credit line.[26] The impetus behind the TOD fund, ULC had as of 2013 acquired TOD sites at eight stations using $11.8 million in TOD fund resources to assemble sites for affordable housing TODs. The **Evans Station Lofts**, the first Denver TOD Fund project, opened in 2013 with fifty affordable units.

- **RTD TOD Strategic Plan 2010**. The plan marked a departure from the passive TOD program RTD had been running since 2000 to one that had become more proactive, seeking outcomes such as encouraging TOD, housing affordability, job creation and the ability to replace some commuter parking for TOD.[27]
- **RTD TOD Pilot Projects 2010**. Following the adoption the Strategic Plan, RTD solicited and received eight letters of interest from local governments to partner on TOD pilots at twelve stations. The pilot proposals were evaluated based on factors such as the ability for RTD to help "nudge" development; having a TOD plan in place; the commitment by local government and the developer/property owner; and, market potential. Four pilot projects were selected – Alameda Station, Old Town Arvada, Federal Center and Welton. The pilots gave RTD a chance to get its hands dirty with real projects and "learn by doing." The pilot at Alameda station was the first to open in 2015, replacing 300 commuter parking spaces with 275 units of market rate housing and a transit plaza.
- **Denver Livability Partnership 2011**.[28] Fueled by a U.S. Department of Housing and Urban Development Community Challenge Grant ($1.8 million) and a U.S. Department of Transportation TIGER Grant ($1.2 million) the effort focused on capacity building, planning for affordable TOD in the West Corridor and funded the **Denver TOD Strategic Plan Update 2014**.[29] The TOD Strategic Plan update assessed lessons learned from the original plan and shifted the focus towards implementation of TOD. The plan identified citywide strategies and station specific strategies for implementing TOD throughout the city.
- **Housing Development Assistance Fund 2011**. An outgrowth of the Livability Partnership, the fund was created to help preserve land for affordable housing within one-half mile of transit stations and high frequency bus routes. Sub-grants were made available to the ULC and to the Denver Housing Authority of up to $750,000 each. These funds ensured that pre-development and holding costs do not get passed through to the renter/owner, thereby lowering housing and transportation costs for West Side families.[30]
- **RTD TOD Assessment 2015**. Pivoting off the 2010 Strategic Plan the RTD Board sought an outside assessment of their TOD program to understand specifically what RTD needed to do to make its TOD program more effective and efficient as it prepared for the opening of five new lines in 2016. In order to move TOD to the next level the assessment found RTD needed to be more systematic, build new processes, be proactive and elevate TOD within the agency.
- **City of Denver TOD Manager 2015**. In September 2015 the City created and hired its first TOD Manager position. The move reflected the growing importance of TOD to the City and the multi-departmental reach necessary to implement TOD. According to Denver's Director of Planning the TOD Managers job focuses on "identifying and acquiring funding, crafting regulations and helping to prioritize projects to meet citywide goals."[31]

Central city riches, suburban focus

As has been the case with many transit systems, the City of Denver may end up as the big winner when it comes to reaping the transformative redevelopment value of transit infrastructure to reshape and revitalize the areas around stations. RTD's rail system focuses on downtown Denver and will ultimately serve the City with forty-one stations.

Figure 4.7 Denver Union Station, light rail station and TOD. Redevelopment of the 43.5 acre Denver Union Station site is creating a new high-density transit-oriented neighborhood in downtown Denver

Source: Photograph by G.B. Arrington.

Having so many stations can be both a blessing and a curse. It can be a blessing because of the scale of the opportunity for city shaping by harnessing the catalytic power of transit in combination with supportive plans and policies. And it can be a curse because it is hard, if not impossible to effectively focus implementation efforts spanning so many stations and political constituencies. While Denver's Updated TOD Strategic Plan set's a framework for the future the proof of the pudding for how Denver's forty-one stations redevelop will be in the eating.

The first major indication of how Denver's transit driven urban redevelopment strategy is doing is Denver Union Station (DUS). DUS opened in May 2014 and the accompanying historic building rehabilitation in July. DUS includes a new twenty-two-bay, underground bus concourse and commuter rail service will arrive in 2016. In addition to the 2004 Denver Union Station Master Plan, the immediate station has been zoned Transit Mixed-Use by the City of Denver, allowing for a wide variety of residential, commercial, and civic uses.

The 43.5-acre DUS redevelopment area is emerging as a major new transit-oriented downtown district and a leading national example of large-scale TOD. According to RTD the 500+ million dollars in public investment has so far spurred over one billion dollars in private TOD projects within the area. The tallest building in the district stands at twenty-one stories. The offices and residential buildings in Denver's newest neighborhood will be soon joined by a 42,000 square foot King Soopers – the cities first downtown grocery store.[32]

With fewer stations to worry about suburban communities face fewer competing priorities. Consequentially they typically have been able to provide more focus and attention to their priority stations. The City of Arvada with three stations opening in 2016 is a case in point. Arvada has been proactively planning for TOD at its three future Gold Line stations since FasTracks was approved in 2004.

Where Denver has between seven to ten downtown rail stations depending on how you count, Arvada is typical of other suburban communities; it has one downtown station – the Old Town Arvada Station. That has meant Arvada could put all of its eggs in one basket when it came to using transit for urban redevelopment. Not surprisingly the City chose to focus its efforts and resources on the Old Town Arvada Station. That commitment was also a factor in RTD selecting the station as the site of one of RTD's four TOD pilot projects. The City, the Arvada Urban Renewal Authority (AURA), and RTD agreed to jointly work together to facilitate a TOD on 8.8 acres of land RTD owned near the station.

Facing budget constraints RTD had planned on 400 surface parking spaces at the Old Town Station on opening day. While RTD had a long-term commitment to build a structured parking garage, Arvada didn't want to see a large surface parking at their downtown station. Through the TOD pilot the City entered into an agreement with RTD to use urban renewal funds to assist with the design and construction of a parking structure and elevator and to jointly offer the site for mixed-use high-density development. Trammell Crow was selected as the developer partner in the summer/fall of 2013 and commenced the master plan visioning process. Trammell Crow's responsibility also included the construction of replacement RTD parking (400 spaces) within a parking structure as well additional parking (up to 300 spaces) to support private development. The developer's initial private development plans included 200 to 450 units, a 110 room hotel, and 30,000 to 50,000 square feet of retail which would ideally include an urban grocery store.

Prospects for the future

With Denver's first TOD nearing its 16th birthday the region is still a relatively new member of the transit urban redevelopment club. And with the imminent opening of five new lines in 2016 Denver's redevelopment future has yet to be written. Collectively the regions new TOD tools and capabilities provide a firm platform to move the Denver Region forward as it prepares to shape the region with transit. And if DUS is the guide, the Denver Region faces bright prospects for the future. Along the way policy makers will need to be reminded the journey for becoming a transit city is a multi-decades undertaking spanning multiple political and real estate market cycles. And that past is prolog. Just as

Table 4.2 Evolution in Denver TOD activities

Year	Activity
TOD 1.0 Getting started	
2000	**RTD TOD Program** – 1 staff person doing education and advocacy
2002	**Blueprint Denver** – integrated plan focuses mixed use at transit stops
2004	**FasTracks** – voters approve 122 mile regional transit expansion
TOD 2.0 Getting plans in place	
2006	**Denver TOD Strategic Plan** – guide for prioritizing TOD by station
2006	**Denver TOD initiative** – $1m to complete TOD, plans 8 stations
2006–2010	**TOD plans** – 40 station area TOD plans underway across the region
TOD 3.0 Implementation focused	
2010	**Denver TOD Fund** – first TOD fund in US capitalized at $13.5 million
2010	**RTD TOD Strategic Plan** – expands TOD program to be more proactive
2010	**RTD TOD Pilot Projects** – 4 TOD pilots selected from 15 proposals
2011	**Livability Partnership** – grants of $3 million for TOD capacity and planning
2013	**Denver TOD Fund** – $11.8 million, acquires 8 sites for affordable TOD
2014	**Updated Denver TOD Strategic Plan** – focuses on implementation
2014	**Denver TOD Fund** – expands to region and capitalized at $24 million
2015	**Alameda Village** – first TOD pilot opens, replaces 300 parking spaces

Table 4.3 Selected timeline of TOD investments

Year	Activity
2000	**CityCenter Englewood** – region's first TOD opens replacing a dead mall
2004	**Denver Union Station Master Plan** – approved for 42.5 acres
2007	**University Lofts** – 5 story 35 unit residential TOD
2008	**Lincoln Station** – 180k mixed-use office and retail
2013	**Mariposa** – Phase 2 of 9 creating a 900 unit mixed-income TOD
2013	**Evans Station Lofts** – first Denver TOD Fund project 50 affordable units
2014	**Lamar Crossing** – first West Line TOD 110 units 80% affordable
2014	**Denver Union Station** opens – more than $1B in TOD completed
2015	**Alameda Village** – first TOD pilot opens replaces 300 parking spaces

they borrowed from other communities, they built new tools to get where they are today. The path to the future will no doubt require redoubling those efforts. Whether the Denver Region will maintain its resolve and will be equal to the opportunity remains to be seen, but the progress to date is encouraging.

Notes

1. http://en.wikipedia.org/wiki/Pacific_Electric.
2. Email from Julie Gustafson, Community Relations Program Manager, Portland Streetcar, Inc. 12.12.14.
3. Public and Private Investments in South Lake Union, July 2012, Berk Consulting and Heartland LLC.
4. http://seattletimes.com/html/localnews/2019227494_amazon22m.html?prmid=4939.
5. Duncan Watry, Internal BART Memorandum to Grace Crunican "*Plan Bay Area Jobs and Housing Projections (2040)*," July 24, 2013.
6. www.bart.gov/sites/default/files/docs/Executive%20Summary%20Building%20a%20Better%20BART.pdf.
7. www.bart.gov/news/articles/2014/news20141009-0.
8. www.fairfaxcounty.gov/dpz/tysonscorner/vision.htm.
9. www.slideshare.net/fairfaxcounty/tysons-board-presentation-all?related=1.
10. www.fairfaxcounty.gov/tysons/implementation/annual_report.htm.
11. Pearl District Access and Circulation Plan Existing Conditions Report. Portland Bureau of Transportation 2009. www.portlandoregon.gov/transportation/article/306707.
12. http://portlandstreetcar.org/node/28.
13. "Transit-Oriented Development in America: Experiences, Challenges, and Prospects," Report 102, Washington, DC: Transit Cooperative Research Program, National Research Council, 2004.
14. Townsend, Anthony, *Smart Cities; Big Data, Civic Hackers, and the Quest for a New Utopia*, W.W. Norton, 2013; Washburn, Alexandro, *The Nature of Urban Design*, Island Press, 2013.
15. RTD, "2014 Transit-Oriented Development Status Report," undated Denver RTD.
16. www.miamidade.gov/citt/library/summit/2015-transportation-summit-presentations/phil-washington-presentation.pdf.
17. G.B. Arrington, "Statewide Transit Oriented Development (TOD) Study: Factors for Success in California. Special Report on Parking and TOD: Challenges and Opportunities," California Department of Transportation, February 2002.
18. Author's personal collection: "Transit Development Planning in the District of Columbia," John Fondersmith, Director Transit Development Team, Office of Planning and Management, District of Columbia Government, October 1971.
19. www.denvergov.org/content/denvergov/en/community-planning-and-development/planning-and-design/blueprint-denver.html.
20. http://ctod.org/pdfs/2006TODStrategicPlanDenver.pdf.

21 www.denvergov.org/Portals/193/documents/40th%20and%2040th/38th%20and%20Blake%20 8_11_09%20for%20web.pdf.
22 http://static.arvada.org/docs/1194983579Transit_Station_Plan.pdf.
23 www.enterprisecommunity.com/servlet/servlet.FileDownload?file=00Pa000000KiJOMEA3.
24 www.enterprisecommunity.com/denver-tod-fund.
25 www.urbanlandc.org/denver-transit-oriented-development-fund/.
26 www.citylab.com/housing/2014/12/part-land-bank-part-community-focused-credit-line/383417/.
27 www.rtd-fastracks.com/media/uploads/main/TODStrategicPlan-final_090210.pdf.
28 www.denvergov.org/content/denvergov/en/transit-oriented-development/denver-livability-partnership.html.
29 www.denvergov.org/Portals/193/documents/DLP/TOD_Plan/TOD_Strategic_Plan_FINAL.pdf.
30 www.denvergov.org/content/dam/denvergov/Portals/193/documents/DLP/DLP_Summary.pdf.
31 http://blogs.denverpost.com/thespot/2015/10/30/ex-councilman-chris-nevitt-lands-city-transit-development-job/123618/#more-123618.
32 Ibid. RTD 2014.

5 Parks, open space, arts and culture

Barry Hersh

The provision of open spaces, parks, plazas, playgrounds, promenades and more, are a key element of cities and especially of urban redevelopment. As redevelopment often means an overall increase in density and intensity of use, the provision of outdoor recreation and natural spaces becomes crucial.

Outstanding examples of design and public spaces abound: Daniel Burnham – Chicago Lake Front; Frederick Law Olmstead – Central Park, Prospect Park and Boston's Emerald Necklace and others from Buffalo to Trenton, NJ, and Bridgeport, CT. Olmstead famously said that "parks are the lungs of the city."[1] Open spaces are far more than just an amenity in urban redevelopment; they define the nature of the project and are a key to long-term sustainability. Determining the open space plan must incorporate not just recreational use and aesthetics, but also place-making, market impact, ecological benefits and long-term management. It is the public realm that defines the character of a community, as well as the buildings.

Figure 5.1 Piedmont urban park, Atlanta, Georgia
Source: Wikicommons.

The provision of adequate open recreation space has been a goal of city planning throughout history, and recent public health and environmental concerns have strengthened that resolve. Planners traditionally measure parks (or open space) per 1,000 residents or what proportion of residents have a recreation space within a quarter mile walk. The most traditional park uses, recreation, playgrounds and ballparks remain important as do places to picnic and relax. There is now more emphasis on providing active recreation and green connections, such as walkways and bikeways that connect communities. The urban canopy, trees, are a vital aspect of urban life, providing shade and reducing the heat island effect while adding beauty.

Bikeways play a unique role in urban redevelopment; they are a form of transportation, personal and family recreation and healthy exercise. Bikeways help link and transform communities but have had to fight their way into transportation and municipal budgets. Including bikeways in road, highway, community and urban redevelopment project plans has become more common. Bikeways, especially those along waterfronts, across bridges and through scenic areas have become a valuable asset improving health and the quality of life, whose inclusion can help urban redevelopment succeed.

The environmental benefits of urban open spaces are numerous; trees and other vegetation provide shade, cooling and convert carbon dioxide to oxygen. In this era of global warming and weather events, the ability of open space to retain moisture and reduce the urban heat island effect is increasingly important, as noted in the US Green Building Council LEED standards. Similarly, the use of swales rather than pipes, dunes and wetland instead of seawalls, providing vegetated green spaces to absorb rain, has become a more important aspect of urban storm water management.

There is a significant human benefit to what is called active design, features that encourage walking and other forms of exercise. Some of these features may be inside buildings such as convenient and highly visible stairways. Many of the park and open space features that have environmental benefits also provide walking and recreation spaces that promote a healthy lifestyle.[2]

The real estate investment community is well aware that, in general, urban land values are higher near parks and open space. People like to live near, look at, shop and dine near open spaces. Both small and large urban redevelopment plans almost always include the provision of new or improved open spaces.

Bryant Park, between Grand Central and Times Square, in New York City is a remarkable urban rebirth story. In the 1970s, the park was dark, dirty and dangerous with drug traffic. William H. Whyte, the noted sociologist observed and filmed the park and provided the key ideas for the parks restoration. The park was redesigned to be more open and accessible, a great lawn was created. Movable chairs that some thought would be stolen proved safe and well-used for many events. Instead of the blank back wall of the main public library, restaurants were put in place with more food choices elsewhere. While some decried "businesses" in the park, the result was more activity, convenience and safety. Winter activities, shops for the holidays and an ice skating rink, were installed seasonally. Park usage soared, and major office buildings including One Bryant Park and Seven Bryant Park have been completed on corners diagonally across, but with views of the park.

Arts and culture

Artists are well known to real estate developers as pioneers and early adopters; when artists and funky galleries move into a downtrodden urban community, it often means that things

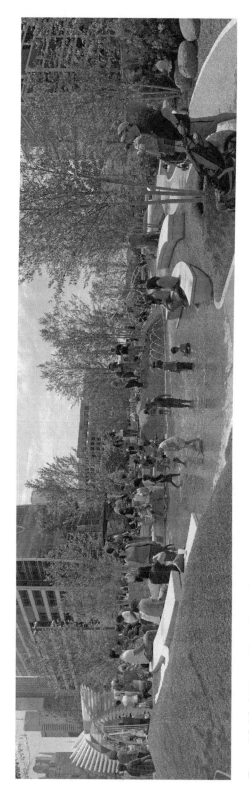

Figure 5.2 Klyde Warren Park, Dallas, Texas
Source: Design and photograph by ElevateArchitecture.

will soon turn around. Everywhere from Deep Ellum in Dallas to DUMBO (Down Under the Manhattan Bridge Overpass) in Brooklyn, NY, and SLU in Seattle, artists have been the first to settle in low-cost urban neighborhoods and start revitalization – and some would say ironically start displacement as well as gentrification.

Community facilities are a key element in urban revitalization, and arts venues, theaters, museums and galleries, artist work-live lofts, are all cultural features that along with parks strongly support urban redevelopment projects. Part of place-making, so important in attracting millennials, is providing places that feel hip and cool, murals and other outdoor art, cafes, dance and yoga studios, black box stages, all contribute to transforming a gritty neighborhood into a desirable urban place with character.

David Bromberg has been known as a unique blue-grass guitar player and singer since the 1960s, performing with Bob Dylan, Jerry Jeff Walker, Jerry Garcia and many other folk and blues legends, as well as with his own groups. More recently, he has branched out, making a movie, performing in a violin and guitar duet, a Grammy-nominated album and most especially as a premier luthier, a maker and repairer of fine stringed instruments. His shop is located in downtown Wilmington, sitting at Market and Sixth streets, across the street from the Delaware College of Art and Design and down the block from the Christina Cultural Arts Center; neighbors include a handful of discount stores, Paradise Caribbean Cuisine, the World Café venue and Coming Soon! Signs, an illustration of the role of the arts in redevelopment and revitalization. Bromberg has participated and brought in others to numerous concerts and venues, including now famous Thursday evening downtown Wilmington jams, bringing fans and helping to build an arts scene. After considerable public investment, several developments including the 380 unit Buccini/Pollin Group project and a former bank, historic adaptive reuses are underway. While not as expansive as Austin, Texas, or Nashville, Tennessee, the shop and the music have contributed significantly, along with numerous more typical redevelopment efforts such as streetscape, "Light up the Queen" and commercial projects, bringing a new, upbeat to an older urban setting.[3]

MASS MoCA, the Massachusetts Museum of Contemporary Art, is an extraordinary example of arts and culture leading urban revitalization. The facility now provides over 100,000 square feet of gallery space in 19 former factory buildings on a 13 acre campus in North Adams, Massachusetts. The site was formerly almost a city within a city for Sprague Electric and earlier built and occupied by Arnold printers. The site was listed on the National Register of Historic Places in 1985 and was also a Superfund site. Conceived in the mid-1980s, in part by nearby Williams College faculty, including Thomas Krens who later became famous as head of the Guggenheim, MASS MoCA opened in 1999, with the help of the state and private supporters. The feasibility study was conducted by several world famous architects and the design, primarily rehabilitation of mostly brick former industrial buildings, was by Bruner Cott & Associates. The large-scale modern art exhibits have become a major attraction, further augmented by festivals and live performances, such as the Bang on a Can Summer Institute. This anything but stuffy museum has brought new life to North Adams, Williamstown and Berkshire County, attracting over 150,000 visitors this past year.

BAM (the Brooklyn Academy of Music) was built in 1861 and long provided a venue for famous speakers, plays and musical shows. By the 1950s, the building and the neighborhood had become run-down. Enter an energetic President, Harvey Lichtenstein, who brought in new, cutting-edge shows, raised funds from public and philanthropic sources and brought new audiences, some from Manhattan. The transformation of BAM became the lynchpin for a series of urban redevelopments, including Metrotech closer to downtown

Brooklyn to the north and Pacific Park (formerly known as Atlantic Yards), which includes the Barclay's Arena, home of the Brooklyn Nets, Islanders and shows ranging from Jay-Z to Barbara Streisand. BAM now has three venues and an annual budget of $54 million, the surrounding urban redevelopments, which include hotels, high-end and high-rise residential, new retail and at least eight other cultural facilities have a total value in excess of $6 billion. Brooklyn, as Marty Markowitz the former Borough President, liked to note was now called "tres chic."

Institutions

Educational and medical institutions, often deeply rooted in their communities, have often played a key role in urban redevelopment. Pittsburgh has based its revitalization upon what real estate people call "Eds and Meds," its outstanding universities and strong health care facilities. Major universities, which often include hospitals, have frequently led major redevelopment efforts; University of Pennsylvania in West Philadelphia, University of Chicago, Yale in New Haven, Connecticut, and Johns Hopkins in Baltimore. Smaller colleges, such as Pratt Institute in Brooklyn, Trinity College in Hartford and Clark University in Worcester have been instrumental in turning around their neighborhoods and encouraging redevelopment.

Inevitably there are conflicts; the institutions' interests such as expansion are not consistent with those of neighbors, frequently low-income and minority communities. Some goals such as improved safety may be shared. So while the institutions offer an economic base of jobs as well as services, any redevelopments is tempered by community concerns. These town-gown conflicts have long occurred, sometimes resulting in litigation and political pressure that can delay, change or even derail urban redevelopment.

Mini-case studies

The following are a series of brief descriptions of notable parks and community facilities that were redeveloped on infill sites in cities of various sizes. Each park illustrates the various types of locations, activities and design character that generally smaller urban parks bring to a community. These redevelopment are generally open to the public and were financed by government support as well as contributions to non-profit sponsors.

Gas Works Park, Seattle

Completed in 1975, Gas Works in Seattle remains an innovative and iconic restoration of a former brownfield into a city park. A former manufactured gas plant site, notorious for their difficult cleanups, the design by Richard Haag chose to retain some of the industrial artifacts, notably towers and machinery used in the industrial process, creating an urban park that some called strange. The boiler house has been converted to a picnic shelter with tables, fire grills and an open area. The former exhauster-compressor building, now a children's play barn, features a maze of brightly painted machinery. Controversial at the time, almost 40 years later, the park is still a reference point for the remediation and creative reuse and redesign of a contaminated industrial site for public open space and recreation. Gas Works Parks is somewhat isolated, abutting Lake Union and a still industrial area, but its impact has been as an icon, of Seattle, but also of imaginative reuse of former industrial facilities.[4]

Dry Gulf Stream restoration at Lamar Station Crossing, Lakewood, Colorado

Lakewood, a suburb of Denver in Colorado, used a combination of green infrastructure, TOD and brownfield restoration all to trigger urban redevelopment. While impact of the LRT link to Denver has been the most noted catalyst, the restoration of Dry Gulch Stream, has also been important. The construction of 176 new affordable green housing units benefits from the improved stream-banks and open space, as well as from the improved access to jobs. As an Arts and Cultural District takes place, the redevelopment of forty brownfields, partially funded through an EPA Brownfield Assessment Grant, helps transform this community

Greenway, Ranson, West Virginia

Ranson, with a population under 5,000, is classified as a city in the state of West Virginia and has suffered plant closings and job losses. The designation of a Green Corridor, including improvements to public parks and green infrastructure, has helped revitalize this community. Ranson and adjoining Charles Town received over $6 million in federal grants, a HUD Brownfield Economic Development Initiative (BEDI), brownfield and sustainable community grants that helped clean and replace six brownfield sites including a former brass factory with a new mixed-use development, Powhatan Place. There is also an EPA area-wide Brownfields Pilot grant that supports redevelopment of a parallel commercial corridor.[5] The recent groundbreaking for a new university facility marks a significant step for this redevelopment effort. The city also enacted a new master plan and zoning ordinance to follow through and implement the Green Corridor concepts.

Myriad Botanical Gardens, Oklahoma City, Oklahoma

Myriad Botanical Gardens can trace its roots (pun intended) to Dean A. McGee (founder of Kerr-McGee), an original concept of I.M. Pei, and a later plan by Conklin + Rossant. Owned by the Myriad Gardens Foundation, the latest rejuvenation was designed by the office of James Burnett and Murase Associates, in a joint investment effort that takes an under-utilized yet prime 17 acre urban downtown garden and park site and turns it into a state-of-the-art, highly active destination that improves the quality of life in Oklahoma City and continues the renaissance of the entire downtown. An outstanding feature is the Crystal Bridge Tropical Conservatory – a 224 foot (68 meter) living plant museum featuring towering palm trees, tropical plants and flowers, waterfalls, and exotic animals. The garden also features extensive educational programming and venues, as well as an extensive and varied plant selection. In 2015, Myriad Gardens was an Urban Land Institute Open Space Award winner.

Durham Performing Arts Center, Durham, North Carolina

Since the 1970s, the American Dance Festival has been held annually at Duke University in Durham. As the festival grew, the city proposed turning the derelict industrial area south of downtown into an arts and entertainment district that would include a performance venue large enough to host the festival. Garfield Traub Development of Dallas, Texas, and Chapel Hill–based architect Szostak Design, which also designed the facility, co-developed the Durham Performing Arts Center for the city.

Figure 5.3 Durham Performing Arts Center, North Carolina
Source: Durham Convention and Visitors Bureau.

The 2,800 seat multiuse theater opened in 2008, playing host to Broadway shows and other touring and locally produced performances, as well as serving as the primary stage for the American Dance Festival. Bridging a railroad right of way, it links downtown to the arts and entertainment district. A multilevel glazed lobby with fritted, insulated glass on three street frontages showcases the activity inside. The plaza outside the theater includes a sculpture by Spanish artist Jaume Plensa.[6]

On a smaller scale, Asheville, in the western portion of North Carolina, long known as an eclectic, funky community, combined arts, brownfields and historic preservation in the River Arts District. Riverlink, a local non-profit, renovated the Historic Cotton Mill that provides studio space and builds the urban fabric through the River Arts District,[7] which also helps grow tourism.

Spruce Street Harbor Park

The Philadelphia Delaware River waterfront was historically industrial and underutilized in recent years. While some commercial uses, such as the Urban Outfitters headquarters at the former Navy Yard, have been successful, much of the waterfront awaits reuse. The Spruce Street Harbor Park is full of creative activities including a lily pad/floating garden built upon three barges and six shipping containers. There is a new beach, boardwalk, net lounge over the water, mist garden and a hammock island. Seven islands, made of recycled plastic, draw water for the plants on several "islands." There are a host of pop-up related food facilities as well as a roller rink sponsored by Blue Cross. This park draws large crowds in the summer, and there are now also winter activities on the waterfront.[8]

Discovery Green, Houston, Texas

Now over 12 acres, Discovery Green near downtown Houston includes a former railway yard brownfield, two former city surface parking lots plus a former strip retail and office center. The project was strongly supported by the then Mayor Bill White as well as a philanthropic partnership that is now the Discovery Green Conservancy. The cleanup featured by choice the removal, rather than just capping, of thousands of tons of contaminated soils, plus three feet of clean fill and was paid for in part by the seller and by the City. Discovery Green is today enormously popular, filled with a wide range of activities and events.[9]

Notes

1 Frederick Law Olmstead and John Muir both the used the phrase, "parks are the lungs of the city," but the term was used earlier in London. www.barrypopik.com/index.php/new_york_city/entry/lungs_of_the_city_central_park/.
2 LOHAS, Lifestyles of Health and Sustainability.
3 www.nytimes.com/2015/02/04/realestate/commercial/wilmington-caters-to-millennials-in-downtown-development.html?_r=0.
4 www.citylivingseattle.com/Content/News/Urban-Dwellings/Article/Gas-Works-Park-Seattle-s-strangest-park-has-an-equally-weird-history/22/169/90414.
5 Brownfields Area-wide Planning Project Fact Sheet Ranson www.epa.gov/sites/production/files/2015-09/documents/awp_ranson_wv.pdf.
6 http://urbanland.uli.org/planning-design/ulx-ten-brownfields-sprout-new-life/.
7 Riverlink http://riverlink.org/learn/about-riverlink/history-of-riverlink/.
8 The Waterfront Center, 2015 Design Awards, David Fierabend, Groundswell Design Group, LLC.
9 Harnick, Peter and Ryan Donahue, "Turning Brownfields into Parks," *Planning*, December, 2011.

6 Environmental issues – brownfields

Barry Hersh

Among the most challenging aspects of urban redevelopment is dealing with the legacy of environmental contamination. Heavily polluted sites generally fall under federal or state Superfund (CERCLA) regulations[1] or the federal RCRA[2] program for industrial waste facilities. The brownfields program, legislatively authorized in 2002[3] as an amendment to Superfund, started earlier as a "pilot" program to promote development of the many sites that were contaminated but were not an imminent threat to public health or the environment that warranted the long, litigious and complex Superfund process. The bulk of potential redevelopment sites including many former industrial facilities are considered "brownfields" defined by the US EPA as "real property, the expansion, redevelopment, or reuse of which may be complicated by the presence or potential presence of a hazardous substance, pollutant, or contaminant."[4]

Many brownfields are located on potentially valuable redevelopment opportunities with locations near in the urban core, on waterfronts and along transportation routes. However, redevelopment efforts usually struggle due to contamination issues. Beginning in the 1990s, EPA officials including Carole Browner, Timothy Fields, Marjorie Bucholtz and Lynda Garczynski created first a pilot program and after the 2002 legislation were able to offer assessment, job training and a rotating loan fund programs to promote brownfield redevelopment. The brownfields program is a unique success, drawing bipartisan support especially from mayors, and a key tool of urban redevelopment in the United States. Fenced vacant former industrial sites with signs saying "contaminated" makes no one happy, especially the surrounding community and the mayors who want that property reused and back on the tax rolls. The EPA Brownfields program has supported assessments at 18,000 properties and estimates roughly 25,000 acres are ready for reuse.[5] The US EPA ACRES database of "tracked sites" includes all EPA and State program, totals over 500,000 properties, with over fifteen million acres, as of 2011; close to one million acres have reached cleanup goals in all programs.[6] While early estimates were of roughly 450,000 brownfield sites in the United States,[7] broader, recent measures of contaminated sites indicate a far greater number of sites that could be considered brownfields. HUD 2005 data indicates five million acres of vacant industrial land in the United States.[8] Environmental Data Resources, a private company that provides environmental information, has from 1990 to 2010, twenty-three million records of property contamination.[9]

Dealing with contamination adds a major layer of complication to urban redevelopment and EPA Brownfield Assessment Grants have been especially important. A recent Bureau of Economic Research study showed a return of more than $17 for every dollar from the brownfields program.[10] Site pollution must be characterized, a remedial plan proposed and approved, usually by a state agency, and then the remediation work completed to the appropriate standard and stigma overcome. This process involves environmental engineers, hydro-geologists, other consultants, as well as environmental attorneys and even specialized environmental insurance experts. The developer must determine the extent of the issue, the cost and, equally important, the time necessary for the remediation and to determine and mitigate potential liabilities. From the community perspective, the focus is on assuring that the remediation is done properly, that contamination be contained during the cleanup and disposed of properly and that the environmental solution be sustainable. The technical issues can be daunting and expensive, such as dealing with hard to remove contaminants including dense non-aqueous phase liquids (DNAPL)s, polychlorinated biphenyls (PCB)s and heavy metals. There is a wide range of solutions ranging from dig and haul, encapsulation, hot spot removal to *in situ* remediation, use of plants and microbes that can remediate pollution over time, and long-term treatment systems. The technology continues to get better; we can test in parts per billion, use radar and other tools to investigate with fewer instances of digging holes or wells and use Triad and other systems to fix problems in the field and in real time. Remediation today is more cost effective, but can still be expensive.

Since the adoption of environmental protection laws in the late 1970s, and especially after the so-called Brownfield Law in 2002,[11] developers, environmental agencies and advocacy groups have all hoped that more of the estimated 500,000 to 1,000,000 brownfield sites[12] would be addressed. For example, after a former gas station has been closed, the major oil company often retains a portion of the responsibility for remediation costs. Petroleum products are ubiquitous, and most cleanups often involve a "tank yank" and removal of leaked material that falls into a predictable range – yet only these gas station sites have converted to uses such as coffee shops, parks and drug stores or incorporated into larger housing and mixed-use projects, as discussed in the detailed case study below.

Larger, former industrial facilities often pose even more complex problems involving more toxic chemicals such as solvents and heavy metals. As noted earlier, brownfields are generally less contaminated than US EPA Superfund (CERCLA) sites or industrial waste facilities that generally fall under RCRA (an older and in some ways more restrictive remediation law).[13] Financial responsibility may involve numerous parties, including older insurance policies, and contamination may have moved across property lines and into groundwater. Some large projects, for example, former auto manufacturing, electrical generating and other plants, can be redeveloped, often with the help of federal or state brownfield programs and tax credits. Examples discussed here include the former Jeep factory in Toledo, Ohio, the former GM plant in Tarrytown, New York, and former refinery in Hercules, California. Even heavily polluted Superfund sites have been remediated and redeveloped successfully.

The brownfield movement has been most critical in the so-called "rust belt," older industrial cities of the Midwest and Northeast United States. However, many other regions such as the east side of San Francisco Bay, Birmingham, Alabama, coal mining regions in Appalachia all have significant numbers of contaminated properties. In communities where the former economic base of industry has deteriorated with only partial replacement, brownfield redevelopment is far more difficult. Brownfield sites in strong markets with a value when clean that far surpasses remediation costs can be attractive to developers. Properties where the remediation cost exceeds the value when clean are referred to as "upside down" and hard to reuse. Those sites in the middle require some form of subsidy; assessment or cleanup grant, tax credit or tax abatement, in order to make redevelopment financially viable. Brownfield revitalization is extraordinarily complex; incorporating real estate economics, land use, community benefits, ecology, hydrology, sustainability metrics, design, politics and a host of associated disciplines. There's also an array of regulatory and funding agencies at federal, state and local levels and often elaborate impact analyses and mitigation strategies. Developer concerns include site analysis, land reuse approvals, market analysis, financing, remediation and liability approaches, project organization and sequencing, waterfront design and shoreline improvements, as well as the host of regulatory reviews involved.

Among the strategies discussed are leadership roles and team building; innovative financing tools including government programs; techniques such as charrettes, checklists and critical paths to aid information flow and support creative planning and design. Specific approaches to difficult aspects such as acquisition strategy and synchronizing remediation and redevelopment are discussed in detail. While there is no silver bullet, there are a set of strategic pathways towards successful redevelopment of waterfront brownfields.

The US EPA "Handbook on the Benefits, Costs and Impacts of Land Cleanup and Reuse" did make full use of EPA ACRE and other data sources and provides the equilibrium analysis below that quantifies land values of brownfield sites in relation to the center, or 100 percent location, in real estate analysis. The same concept could be applied by measuring distance from waterfront or any other desirable real estate center. The "dip" in the center of the graph illustrates a decline to no value for a property with known, severe contamination. As the remediation occurs and is documented, the value returns to the same pattern. This is, on the ground, exactly what redevelopers try to do, restore a brownfield so that it becomes and is valued as conventional real estate.

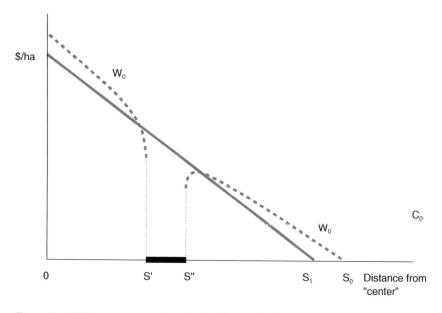

Figure 6.1 Willingness to pay for urban land after site cleanup: general equilibrium analysis
Source: United States Environmental Protection Agency, David Lloyd, Brownfield Program Director, citing Bureau of Economic Research Brownfield Study, December 2011.

The general equilibrium benefits measure, represented by the dotted area in this figure is smaller than the partial equilibrium benefits measure in (a previous chart) because of this decrease in land values throughout the market. Of course, this benefits measure is still positive even if equilibrium property prices throughout the urban land market fall relative to the baseline scenario. The fall in prices at properties not directly affected by the cleanup, represented by the two dashed areas in this figure, leads to a transfer of resources away from current property owners. Some of the transfer would go to current renters and the remainder to future buyers. The dashed areas represent a pecuniary effect rather than a social benefit or cost of the land cleanup program.

This example illustrates a more general result that estimates based on property values that do not account for general equilibrium adjustments can overestimate the benefits of land cleanup and reuse. The size of the discrepancy between the two estimates depends on the quantity and importance of the remediated land compared to the size of the property market as a whole. Cleanup programs that target very large sites or many parcels throughout a real estate market are more likely to have far-reaching effects on prices than programs targeting smaller or fewer sites. Equilibrium effects are also more likely if residents cannot easily move between cities, or if cleanups occur in multiple cities simultaneously, both of which raise the total quantity of remediated land relative to the area over which people make decisions about where to live.

Changes in prices throughout a real estate market make it more likely that residents decide to move as a result of the cleanup. For instance, an improved appearance and lower health risks that are capitalized into higher housing prices near a cleaned up site could spur some renters to move away in search of cheaper housing, while other residents who previously avoided the neighborhood due to the contamination move closer. Such spatial sorting can complicate the empirical analysis of land cleanup benefits using property values, as will be discussed further in a further section. Evidence has been found of residential sorting in response to improved air.[14]

The statistics and graphs shown demonstrate the value and issues of quantitative analyses, whether using hedonic or equilibrium techniques, based on the data available. Numerous studies, including those by Professor Peter Meyer of the University of Louisville, have assessed the positive impact of brownfield redevelopment on distressed urban areas.[15] There have been a number of recent university studies that support brownfield redevelopment as an effective tool in improving property values and creating jobs.[16] While it is difficult to separately analyze total brownfields and to attempt to segment those on waterfronts, it is anecdotally clear there are significant numbers of properties with some level of contamination, and a relatively smaller number of sites that have been remediated, and that substantial waterfront brownfield opportunities remain. The analyses do indicate that contaminated sites, even waterfronts, may essentially fall off the table in terms of value when significant contamination is first found – but can return to the overall pattern after remediation and redevelopment. There needs to be both the reality and the perception that the specific location is safe.

That a brownfield is defined as property where redevelopment is "complicated" not only results in a large number of brownfield opportunities, but also points to the challenges to restoring value to a waterfront brownfield property. Real estate development is inherently a complex and risky endeavor, involving site selection, design, land use planning, market feasibility, financing and more. Working on brownfields adds an additional layer of complexity: assessing site contamination, remediation method selection and, perhaps most importantly, environmental liability issues. Waterfronts have their own layer of concerns: preservation of maritime activities, shoreline access and treatment, ecological concerns, design constraints and opportunities, to name a few. Finally, brownfield redevelopment has inherent sustainability features; for example, LEED certification for both buildings and neighborhoods, recognize and value brownfield remediation – but there are further steps, especially related to energy conservation and storm water design, that are necessary for recognition as "green." Gaining the full value from an environmentally distressed asset is a challenge indeed.

The community perspective on brownfield remediation has been said to resemble the stages of dealing with a major illness. First is denial; our community is not "dirty." Then there is hoping that it is not so bad and will go away. Then there is anger, at the polluters and the government for allowing the pollution – just send in trucks and take every molecule somewhere else. Finally, there is acceptance, negotiating what is the most beneficial solution, one that truly protects the community as soon as possible, during remediation and construction and for the long term. All this is in addition to the community role in any development: concerns about land use, community character, traffic and gentrification.

Many contaminated sites including some of the largest and best known brownfield projects are on waterfronts, adding another layer of technical and bureaucratic complexity. Many cities, including New York, Charleston, Baltimore and Oakland have worked for decades to rediscover their waterfronts. These redevelopments are creating promenades, shopping and restaurants where shipping piers and factories once stood. These projects involve a range of environmental concerns in addition to remediation of contamination, including design for storm water runoff and flood resiliency, restoration of wetlands, improving marinas and other water dependent uses, as well as public access to the waterfront.

Brownfield projects are often the signature projects of urban revitalization: baseball parks in Trenton and Bridgeport, food markets in Seattle and Boston and other urban placemaking. These sites often have significant advantages, such as being within walking distance of transit hubs or on waterfronts. The obstacles can also be significant, including the need to deal with multiple regulatory agencies and technical remediation choices that range from straightforward "dig and haul" to innovative techniques that prevent contaminants from migrating from one property to another or between land and water.

Often a public-partnership is advantageous to help coordinate between land use and economic development and to provide financial support for regional problems such the need to dredge a waterway. While often involving a public-private partnership, these well located properties, are often walkable to transit and job centers, still need a combination of public as well as private capital and enormous staying power to deal with the plethora of agencies, including US EPA, state environmental regulatory agencies, municipal economic development and land use agencies. There are many successful brownfield redevelopments in every state, many of which are listed as EPA Phoenix Award winners[17] or Brownfield Renewal recipients.[18] Technology of testing and remediation is steadily improving and cleanups have become more cost effective. Over the last few years, developers, lenders and regulators have gained valuable experience, and as a result, there is less of a stigma associated with formerly polluted sites. Despite this progress, the overriding brownfield concerns are how to control costs and accelerate the process, so contamination issues do not delay or even kill urban redevelopment opportunities.

Private investment and development of brownfields requires a level of strength in the local real estate market; in particular, lenders remain cautious, trying to focus on the relatively few remaining opportunities with high real estate potential and manageable remediation. There remain many properties with high remediation costs and relatively little real estate value. Often property owners and potential developer buyers hit an immediate snag as to who will pay for site environmental assessment and who will deal with what is found. If the problem is manageable, the developer may proceed; but in the minority of cases where the problem is larger than anticipated, the owner is now responsible as the developer will drop the project. This is why assessment grants, as earlier noted, are especially useful in resolving initial site characterization issues, thus leading to successful brownfields redevelopment. Such early stage assistance and understanding the individual site issue are often the catalyst and have proven so cost effective.[19] Among the many communities that have benefited from these EPA assessment funds is Trenton. The new waterfront park baseball stadium project and the Crane industrial redevelopment were started with such assessment funds. Very often once a site has been characterized, a remediation plan can be determined, a budget established and funding from various sources (responsible parties, the owner, government agencies, lenders and the developer) lined up. A related technique is to try to plan early and carefully to gain some synergy between remediation and development work (i.e., having one contractor do cleanup and development site work) so as to gain both time and efficiency

Conversely, tax abatements for remediation are generally paid after cleanup and when redevelopment is complete. This may have the advantage of being off-budget for government agencies and have the sureness of only being paid after cleanup completion. The drawback of this approach is that all the assessment and remediation must be paid from other sources, with the expectation of receiving valuable tax benefits only at the end of the process. Brownfield tax credits, unlike historic and low-income housing tax credits are only available in some states and have not established a market where they can be sold and thereby provide equity.

Environmental insurance plays an important role in brownfield and other urban redevelopment as a tool to manage environmental liability. There are various forms of environmental insurance offered by numerous insurance companies, providing various forms of protection such as from unknown contaminants that appear during construction despite careful due diligence or from claims made by other properties or former users of the property. Some forms of environmental insurance can help protect against unexpected or larger than expected remediation costs. Environmental insurance has been described as the grease in the deal, in particular giving comforts to deep-pocket participants such corporations selling property and institutional lenders. There are highly specialized insurance company groups, as well as insurance brokers and environmental lawyers whose expertise helps make environmental insurance work effectively.

Another persistent question is how to do "brownfields by the bunch"[20] to improve productivity, reducing the time and cost of individual projects. This can mean privately owned portfolios, such as gas station properties as previously mentioned. Shopping centers are also often considered brownfields due to tire-battery-accessory operations and dry cleaners. For example, a Canadian portfolio of K-Marts was sold to Cherokee, then the largest brownfield redeveloper in North America. From a policy perspective, government often can provide infrastructure and parks, which set the stage and incentivize within available government resources.

Cleaning up an entire community also involves improved water quality. In large part due to the Clean Water Act of 1972[21] as well as related state, federal and Canadian laws, potable water has been significantly improved and assured in many urban areas. Urban redevelopments are often among the beneficiaries of clean, inexpensive municipal water. Remediation and redevelopment often includes costly steps to protect groundwater and especially any nearby aquifers.

The area-wide approach also deals with multiple brownfield sites and is usually reserved for low-income neighborhoods. This community scale approach aims at starting neighborhood revitalization, including several brownfield sites, perhaps coordinating cleanups, and often ties into other community improvements and TODs. These plans address multiple issues, such as reusing brownfields into needed parks or community facilities as well as residential and commercial real estate projects. New York, with its Brownfield Opportunity Area planning grants, was an early adopter, but this program has been cumbersome and there new efforts revive. The federal effort, tied to the EPA-HUD-DOT Sustainable Communities effort has seemed to have gained some traction, in part because of the multi-agency participation and ties to TOD and place-making. Redeveloping brownfields is seen as smart growth, an alternative to suburban expansion, or sprawl.[22] For a time after the state and federal governments initiated brownfield programs and traditional urban renewal programs lost favor, brownfields were the only new money and energy in community revitalization. In the past few years, brownfields have to some degree been subsumed into the broader efforts towards overall sustainability and resilience in urban redevelopment. Not only can the brownfields remediation be seen as an improvement in overall environmental quality, even the remediation itself is now designed to be energy efficient and limit even temporary environmental issues.

Remediating contamination has been seen as turning a polluted property into a conventional real estate opportunity. Ideally, there can be a synergy between remediation and redevelopment, the coordination of actual construction and of financial flows, which maximizes efficiency and benefits the project. Today, brownfields remediation is still a challenge, but it is part of improving the overall environmental quality of an urban environment and creating successful city redevelopments.

Other environmental concerns: noise and air quality

One of the most common complaints about urban living is traffic, construction and electronic devices all contribute to noise levels. Most large cities have noise ordinances that limit the decibel level that can be generated, usually in nighttime hours. There are both planning and design approaches to limit noise. Residential streets can be planned to reduce through and truck traffic, limiting noise in residential neighborhoods. Hours of operation for bars and concerts can also limit noise. Buildings can be designed with noise limiting windows and HVAC systems to insure quiet within residential units. Limiting noise is a legitimate factor to be considered in the planning and architecture of urban redevelopments.

Air quality is another regional factor that impacts the desirability of urban living, but it is harder to control locally. Cities generally have benefited from air emission restrictions on vehicles, power plants and other structures, so that air quality has been improved, though it varies from region to region. Still many high-performance buildings claim that the air inside their building is better than the outside air – and take measurements to demonstrate the difference. Air intakes can be located as high as possible (generally lower particulant matter and some other pollutants) and various types of filters and scrubbers can be used to improve indoor air quality.

Waterfront redevelopment

Barry Hersh, updated and revised from **Urban Waterfront Redevelopment, NAIOP, 2012**

There's been a remarkable renaissance on America's urban waterways. The mixed-use redevelopment of formerly contaminated waterfronts has become an important but challenging part of urban revitalization, and also a significant real estate opportunity. The decline in industrial use has led to the opening up of waterfronts for increased residential, recreational and commercial use. In an era when traditional suburban development has become difficult due to transportation costs, environmental concerns and market shifts, in-city waterfront brownfields have often shown themselves to be significant opportunities. While there have been outstanding projects completed over time, the challenge is to provide a framework, so waterfront revitalization can be expedited and made more common, with greater emphasis on long-term sustainability.

Waterfront brownfield revitalization is extraordinarily complex, incorporating real estate economics, land use, community benefits, ecology, hydrology, sustainability metrics, design, and politics across a variety of associated disciplines. Also involved are an array of regulatory and funding agencies, at the federal, state and local levels, and often elaborate impact analyses and mitigation strategies must also be employed. Developer concerns such as site analysis, land reuse approvals, market analyses, financing, remediation and liability approaches, project organization and sequencing, waterfront design and shoreline improvements, as well as a host of regulatory reviews are all involved. The mixed-use redevelopment of formerly contaminated waterfronts has become an important but also challenging part of urban revitalization – and a significant real estate opportunity.

Waterfronts have their own layer of concerns including:

- preservation of maritime activities
- shoreline access and treatment
- flooding and resilience
- ecological concerns
- design constraints and opportunities.

Brownfield redevelopment has inherent sustainability features. For example, LEED certification for both buildings and neighborhoods recognize and value brownfield remediation. However, there are further steps, especially related to energy conservation and storm water design, necessary for recognition as "green." Waterfronts are a great opportunity, but they are even more complicated than other redevelopments. Perhaps most disconcerting to traditional developers, a sustainable waterfront brownfield project is subject to whole new sets of regulators. The following chart illustrates the major categories of review, and the regulator at the federal, state and local levels.

The developer must pick his way through this entire array of reviews in order to successfully maximize the value of the redevelopment. The case studies that follow illustrate how that process can work and how sometimes projects fail and what can be learned from both.

One of the attributes of a complex system is that there are no straightforward rules, only approaches that coordinate numerous aspects of the project, work within that context and

Table 6.1 Waterfront brownfields review matrix

Level	Conventional	Brownfield/ environmental	Waterfront	Green
Federal	FHA, FNMA, Freddie Mac Tax Credit (Low Income, Historic and New Market) via State, CDBG	EPA – CERCLA Brownfield Assess and Tax Credit Endangered Species	Corps of Engineers, Costal Zone, Flood Plain, Wetlands, Ports, Fish and Wildlife	EnergyStar Renewable Energy
State	Environmental Impact Statement, Major Traffic Issues	Voluntary Cleanup Programs Brownfield Incentive	SPDES Discharge Shoreline Design Water Transport Wetlands	Energy Incentives
Municipal	Land Use Permits, Traffic and Parking, Transit Oriented Design, IRBs, Tax Increment Finance	Brownfield Incentives	Public Access Wetlands Storm Water	Weatherization, Green Codes, Storm Water
Private non profit organizations	EIS Comments, Public Hearings	Environmental Justice and Stewardship	Environmental Advocates	LEED
Financial	Equity – Partners Debt – Lenders	Environmental Insurance	Insurance	Green Funds

Legend
CDBG – Community Development Block Grant.
IRB – Industrial Revenue Bond.
CERCLA – Comprehensive Environmental Response, Compensation, and Liability Act, 1980 (Superfund Law).
SPDES – State Pollution Discharge Elimination System (New York and similar in other states).

seem to have broader applicability with adaptation. The following is a set of analytic themes that are aimed at helping to bring order to a level of complexity that can seem to approach chaos, but which developers overcome in order to succeed.

A Leadership and building a team

Real estate development never just happens; leadership that pushes the project forward is always required. Redevelopments, especially waterfronts and brownfields, demand such a key leader, a champion if you will, who is fully committed, can see the broad scope of issues and is both persistent and flexible. Developers are most often that key leader in taking on these complex projects, but mayors, planners and community leaders may be that long-term champion. Experience is very helpful, but repeating what has worked in the past may not be sufficient to find each project's unique path to success in all the issues. Sufficient resources are important; the flexibility and the ability to deal with extraordinary complexity are essential.

J. Brian O'Neill stands out as a regional developer who has successfully focused on a series of brownfield redevelopments. In Harbor Point, the BLT (Building and Land Technology) leadership is clearly the Kuehner family, but John Freeman, as Executive Vice President and General Counsel, is essentially the public face of the project. In Toledo,

Mayor Michael Bell took it upon himself to find a way to finance an important but stalled project. In Portland, the corporate owner's representative, Christopher Grace of Reynolds Metals Development Company, drove the process and found the user. Both Trenton and Stamford had strong planning leadership; it is not coincidental that both were among the first to receive EPA Brownfield assessment grants and had updated master plans – before the developers appeared. Each developer must find the right role for himself or herself and for the project team members.

These complex projects also require not just one but usually several public-private efforts, whether formal partnerships or regulatory. Having relationships with elected officials, both executives and legislators can be crucial – as Dannell Malloy's support for Harbor Point as Mayor and Governor. But mid-level relationships with environmental and land use regulators can also be critical. Managing those relationships, having the right team members from architects to hydrologists, makes all the difference.

Building a team: The team required for a waterfront brownfield is larger and has a broader range of skills than most developments. In addition to the civil engineer, there will be environmental engineers, hydrologists, sustainable storm water designers and possibly other specialists. The legal team will need to be expert not only in real estate transactions and land use approvals but also in dealing with environmental regulators and liability protection. An insurance broker specializing in environmental policies is often used. The architectural team will not only have to deal with building design but also waterfront issues, green certifications and possibly historic preservation. And the developer gets to pay for all of them.

The importance and potential of designing a waterfront is special; standard "big box" or other cookie cutter plans maximize opposition rather than profits. Views of the water, access, the unique history and feel of waterfronts are extraordinarily valuable. Successful projects bring the community back to its waterfront. A designer with the expertise to maximize that value, to coordinate with the coastal and remediation requirements, can be the key to a successful project. The Waterfront Center Award winners, www.waterfrontcenter.com, offer a worldwide set of examples, ranging from major redevelopments to small projects.

As with other aspects of waterfront brownfield redevelopment, each component is important, as is the interaction between each part. So finding, hiring, supervising and incentivizing a team that has the skill, the creativity and can function effectively together becomes one of the developer's most important roles.

B Approval strategies

Waterfront brownfields offer special opportunities for developers. By cleaning up a property, reopening the waterfront to the public and building a sustainable project, the developer has the unusual chance to wear the white hat. Many communities will be hungry for, or at least open to, this type of investment and activity. At some sites, especially those with transit capabilities, greater intensity of use may be allowed than in other parts of the city. One key note is that cities such as Trenton and Kansas City that had dedicated brownfield specialists tended to be more responsive and effective in dealing with the complexity of brownfields and their approval process.

The environmental review process, whether done under the National Environmental Policy Act, signed by President Nixon in 1970, or various federal and state environmental review laws have become well known as a lengthy process, sometimes stretching for not just years but decades. In 2011, the White House Council on Environmental Quality initiated a pilot program to employ "innovative approaches to completing environmental reviews more efficiently and effectively."[23] As noted, innovative programs range from empowering licensed environmental professionals to certify remediation, to presumptive remedies for area-wide issues, and efforts such as New York City's new brownfields program for lightly contaminated sites that have turnaround times measured in days. What is clear from the developer's perspective is that finding an approach that gets to an appropriate, safe remediation in a relatively short time frame is crucial.

The use of the latest technologies, such as sophisticated Geographic Information Systems (GIS) and three-dimensional modeling, improve not only project management but also project communication. Allowing everyone, including agencies and community organizations, to see and understand what is going on with the remediation and the redevelopment.

Some successful case study projects did everything to move ahead quickly agreeing to land use or remediation requests, even at additional cost, in order to move ahead. On large projects, such as Stamford's Harbor Point, this appears to have helped. It's not that the developers did not negotiate hard, but they consistently opted for a viable settlement rather than spend time on additional negotiation and approvals. Being brought in as the fourth developer, they succeeded in hitting the ground running, moving construction quickly and creating a momentum for the project that were supported by early residential and retail success. Similarly, a negotiated, accelerated planning approval helped get FedEx on to the former aluminum plant in Oregon.

The importance of accelerating the land use and environmental approval process can be significant. Projects build momentum, political support and market awareness; slowing down is almost always a negative. Trenton's advance planning allowed projects to move relatively quickly. Stamford's Harbor Point expert team utilized a series of land use strategies, a combination of entitlement following existing provisions in innovative ways, sometimes negotiated specific code revisions and dealt with a number of land use boards. In Troutdale and elsewhere, developers have been able to, in effect, write specific zoning provisions to move a project forward.

That means knowing the limits including innovative government efforts, selecting consultants and attorneys who share the goal of a quick resolution and not getting tied up fighting over details. A sustainable project, one that meets LEED or other standards, may be advantageous. Going from brown to green has appeal to stakeholders, as well as some tenants and lenders.

Stakeholders

Waterfront brownfields involve a complex set of participants, some of whom most real estate developers are not familiar with. Transportation issues may involve not just cars, trucks, buses and trains but also ferries. Boaters and fisherman have specific concerns. Some environmental advocates are concerned about all developments; a waterfront brownfield will bring out different organizations concerned about the remediation, water quality and public access. The community may well have environmental justice concerns about the nature of the cleanup and the type of facilities to be included and the jobs generated. Of course, all the normal land use issues must be addressed. Do not ever think that a waterfront brownfield will quietly proceed, even the projects that had strong support and limited opposition still attracted headlines and blogs. The developer cannot just rely on traditional supporters and contacts and rather must reach out to these disparate groups and address their concerns as early in the process as possible and use new media: websites, blogs as well conventional public relations. Face-to-face contacts and the use of charrettes in early stages can help. Bear in mind that each constituency has contacts with elected officials and reviewing agencies, such as the Department of Environmental Protection and Coastal Zone Authorities. The overall sustainability of the project, building certifications, infrastructure such as storm water controls and the sufficiency of the cleanup may help if done and communicated properly.

C Innovative financing

The complexity of waterfront brownfield redevelopment is reflected in finance, typical private developer financing, mostly equity for site acquisition and upfront development costs, construction loan and a permanent mortgage taken out upon completion is not a sufficiently applicable formula, even adding mezzanine debt or preferred equity. The risks are too great, financing is needed earlier and for longer time periods and there is often a gap between what conventional private financing, both debt and equity, can provide and what the project multiple stakeholders require. Clearly under-capitalized projects are at a severe disadvantage. That being said, there have been numerous examples of both large and small waterfront brownfield success stories. The following are financial mechanisms that have helped waterfront brownfield projects succeed.

- Assessment funding: The first step may be the hardest – who will pay for the initial environmental assessment? Often neither the property owner nor the prospective developer is willing to take that first step, an environmental assessment that will allow a realistic evaluation of the costs and time required for remediation and may constrain redevelopment. The US EPA Brownfields program, along with numerous state programs, provide assessment dollars; EPA alone has provided just over 2,000 assessment grants, usually $200,000 for a total of approximately $480 million.
- The HUD Brownfield Economic Development Initiative (BEDI), which is essentially a loan guarantee program for municipalities to support major brownfield projects, has been effectively used by quite a few developers. This program may be consolidated into other HUD programs in the 2012 budget year. Similarly, the Sustainability Community Initiative grants, which started in 2010, are the subject of annual federal budget negotiations.
- Land acquisition costs: A strategy of many brownfield developers is to acquire property cheaply because of contamination concerns. In some cases, the owner is more interested in avoiding environmental liability than in the sale price. As noted below, it may behoove a developer to take title and control of the property and the cleanup.

Assembling land for urban renewal has been a government role for sixty years, and some brownfield developers have obtained properties at a minimal purchase price – but usually in exchange for dealing with remediation or taking on an important but risky economic development project. The Small Business Liability Relief and Brownfields Revitalization Act of 2002 provided additional and considerable protections for municipalities that take title of brownfields. Today, governments are more likely to acquire brownfields, including waterfront assets, by negotiated acquisition or tax foreclosure – again owners most interested in avoiding cleanup costs and liability – rather than by eminent domain. It is interesting to note that the infamous *Kelo* v. *New London* "taking" case did not much involve environmental issues – but in 2011, US EPA provided an assessment grant for a portion of the subject Fort Trumbull redevelopment. In New York City, a relatively rare exercise of eminent domain recently broke ground on the Willet's Point project. That condemnation, thus far upheld by state courts, was based significantly on long-standing contamination, flooding, infrastructure and other environmental concerns. While not impossible, condemnation can be a slow, politically risky and litigious route; developers and municipalities seem generally to utilize other tools to assemble properties for redevelopment. Dealing with RFPs or other procedures to work with governments and non-profits involving site assembly is discussed in Acquisition Strategies.

- Building financing: In general, government funding for actual construction is tied to the end use, i.e., industrial revenue bond are based upon the business to be relocated, rather than the developer. There are a host of tools available for redevelopment in general, particularly if located in a targeted community. Residential projects may be eligible for low-income housing tax credits and other various housing subsidy programs. Similarly, commercial projects may receive New Market Tax Credit benefits. Waterfronts are often the older part of a city. If the project involves building rehabilitation, it may be eligible for historic tax credits.
- Remediation assistance: In general, it will not be provided to a party potentially responsible for the contamination, but may go to an "innocent purchaser" who has done "all appropriate inquiry" before acquiring the property.[24] Environmental justice and community factors are important in the allocation of remediation assistance at both the federal and state level.

Under the 2002 federal brownfields law, a taxpayer may fully deduct the costs of environmental cleanup in the year the costs were incurred (called "expensing"), rather than spreading the costs over a period of years ("capitalizing"). Unfortunately, only a relatively small number of developers have found this provision sufficiently attractive to utilize.[25] This provision has technically expired but may be extended by Congress as part of the final budget legislation.

A number of states, including New Jersey, Pennsylvania and Michigan, reimburse part of brownfield remediation costs when the environmental regulators have certified the cleanup.

- Waterfront assistance: Federal and state coastal zone management programs may provide assistance – as well as permitting requirements – for planning in coastal zone communities, and there is often additional assistance to retain maritime businesses and facilities – so-called water dependent uses. In addition, there are other categories of support for waterfront amenities, including parks and promenades.
- Corridors and area-wide planning: An increasingly useful approach is to focus on not just one property, but an area or corridor, sometime along a shoreline or greenway. These efforts are led by the HUD-DOT-EPA Partnership, which provides support for housing, community revitalization, transportation improvements, economic development and environmental improvements; either through Sustainable Community Initiatives or various existing federal programs. New York State's Brownfield Opportunity Areas, legislated in 2003, was an early model, and provides planning grants and preference for other funding in communities with "brownfields by the bunch"[26] and

administered by the same agency that supervises coastal zone efforts and state environmental quality reviews. Numerous other states now have similar programs, such as New Jersey's Brownfield Development Areas, which includes and is now assisting 31 communities including Camden, Pennsauken and Trenton along the Delaware River. The area-wide approach supports redevelopment of areas that have a number of relatively "low priority" sites such as gas stations and dry cleaners, but which can benefit from a coordinated cleanup effort.

D Strategies: site acquisition

Finding the right urban development opportunity is the first challenge. Some of the best projects are located in cities, or neighborhoods within cities, that have not seen a great deal of private development in recent years. From Trenton to Oakland, waterfronts in troubled communities are being redeveloped. Developers seeking sites have to go far beyond the obvious, or "me too" sites, and look at the fundamental assets: waterfront, access, infrastructure, market demographics and a community that is open to revitalization. As Tom Darden of Cherokee noted post-recession, there is less interest in "aircraft-carrier" size projects[27] and more focus on projects with a shorter time horizon.

Developers generally seek to control of property for as little upfront cost as possible. Use of options, purchase and sale agreements with long due diligence periods, refundable deposits and contingencies are preferred. Looking at waterfront brownfields or similar complex redevelopment suggests rethinking acquisition strategy.

While actually acquiring property is generally more expensive up front, it comes with several advantages and has become more common for these complex projects. Often a developer, who understands the environmental issues and liability, can negotiate a lower purchase price by buying the property quickly. As noted earlier, there may be government assistance to support assessment and reduce the front end risk of losing substantial dollars on investigation and design. The number of agencies involved, the interplay between them, tends to draw out the process and go beyond what seemed to be reasonable expectations of an option period. Economic and political cycles may intervene. Any rights that have a termination date encourages opponents to stall the project and put pressure on the property owner to kill the deal. This was part of the strategy that foiled Cherokee's vision for Camden and Pennsauken, while some more successful projects, such as Harbor Point, enjoyed the expensive advantage of owning the property as it moved through the process. An owner is taken more seriously by regulators, government officials, neighbors and prospective tenants.

Another category of alternatives is to consider a joint venture with a property owner, which may avoid the expiration risk of an option, but shares the long-term upside in the development. Of course much depends upon the private seller's choice, an immediate gain or a later, more risky, higher price.

In some cases, control of the property rests with a governmental or non-profit agency, so the joint venture means securing control via some form of public-private development agreement, often following an RFP and sometimes performance benchmarks. These make the government entity effectively a partner in the project which may mean a powerful ally, but one whose interest in jobs, open space, environmental protection and design may differ from the developer's interest in return on investment. Public-private partnerships come with their own set of risks; the City of Toledo ended a long-standing development agreement with minimal investment when a new mayor found interest from a Chinese company willing to actually buy the property. If there is an RFP selection process, the criteria are likely to be quite broad; including concerns such as public access to waterfront and amenities provided, as well as affordable housing, the type of retail encouraged and design features, such as not "walling off" water views by maintaining viewsheds from the existing community.

While there are advantages of ownership, developers understand there are very real risks. Actual site acquisition can leave a developer "land poor" – owning potential valuable property but short on cash or over-committed to public amenities. Lenders vary in their willingness to fund land purchases or construction of an environmentally challenged parcel. Looking at the complex waterfront brownfield projects suggests that the likelihood of success is greater with firmer property control, whether actual acquisition, or a contractual commitment as part of a public-private development agreement.

E Synergy between remediation and redevelopment

The intuitive approach is first you clean it up and then you build, often the first thought of neighbors, regulators and most developers – but it is often inefficient. At more than one project, a newly installed remediation cap, was soon broken to install utilities and foundations. In terms of management of complex systems, can two complicated tasks, remediation and construction, be done in parallel rather than in series? Information on remediation informs the plan; i.e., put parking lot over the hot spot and day care center in the clean area. There are ways to coordinate the site remediation and the site design, using the shoreline treatment, whether riprap, sheet pile or natural as a design element, allow public access and simultaneously support the site remediation/encapsulation. In some cases, as in Harbor Point and Troutdale, the remediation and redevelopment site work may be done by the same construction contractor.

Brownfield remediation has significant environmental benefits including: removing hazards to public health, reducing sprawl by environmentally responsible infill development, air quality improvements by reducing both traffic and reduced release of methane and other gasses, improving storm water runoff. A proper remediation is crucial to successfully marketing the property to users, lenders, insurers, and public officials. Developers need to get their arms around the "how much" and "when" of remediation to prepare a realistic pro forma analysis. "All Appropriate Inquiry" a legally defined term, requires sufficient due diligence to assure that the developer is treated as an "innocent purchaser" rather than a polluter.[28]

Estimating remediation costs is especially tricky, each site being unique and complicated by coordination with redevelopment. The Northeast-Midwest Institute in 2008 estimated remediation costs at non-petroleum sites to be between $600,000 and $1,000,000.[29] The US GAO using EPA estimates stated that average petroleum sites, mostly former gas stations with leaking underground storage tanks, cost approximately $125,000 to remediate.[30] Average costs must be considered carefully, there are very expensive outliers, especially sites that involve off-site, groundwater, PCBs (polychlorinated biphenyls) and DNAPL (dense non-aqueous phase liquid) contamination. Most developers are justifiably cautious, but several case study projects, including the Crane site in Trenton, illustrate that remediation cost may be less than first anticipated. New technology and sophisticated environmental testing help make remediation more efficient and less risky.

Just as developers are sophisticated clients of architects and lawyers, it is important for the waterfront brownfield developer to carefully select the remediation team that meets the project's needs and that has sufficient understanding of the remediation/redevelopment process to make certain that the most beneficial approaches are being utilized. A sound remediation plan must be based on robust and accurate data. The remediation/redevelopment plan needs to make effective use of various techniques that help accelerate the process, such as these brownfield-specific proposals:

- Self-certification: Several states including Massachusetts, Connecticut, Ohio and New Jersey have licensed environmental professionals that are allowed to effectively self-certify routine cleanups subject to audit. This can substantially shorten the review and remediation process and should be used when possible.
- Presumptive remedies: Most states have some form of presumptive remedies, areas where the nature of the contamination has been established and there are a set of specific guidelines for cleanup. Where presumptive remedies have been established, as they sometimes have for waterfront corridors, they can help accelerate the design, remediation and redevelopment.
- Triad: is an approach encouraged by US EPA for the past decade and that deals with managing uncertainty in a manner that relates to the theme of this report. Triad coordinates site investigation with remediation, so that a conceptual remediation program is agreed upon, and as work proceeds and data gaps can be identified, new information can be ascertained in the field, and the remediation refined in real time, in the field and in accordance with standards and commitments to stakeholders. Triad remediation tools were used to expedite remediation in both Stamford and Trenton.
- New technologies: are constantly being created in environmental remediation, and because no one solution fits all situations, a sophisticated approach that selects the most appropriate and efficient remedy for the specific situation is needed. Some of the new technologies include bio-remediation and phyto-remediation (as in Trenton), which utilize natural organisms, bacteria, bugs, plants or fungi, to reduce or eliminate contamination, such as the mustard plants used in Trenton. These natural solutions are often desired as they reduce the impacts and energy use of remediation; however, they tend to be relatively slow and less predictable than engineered solutions.
- Environmental liability protections and insurance: The risk of litigation and costs due to potential environmental liability frightens many potential developers and lenders. The environmental liability protections in so-called "no further action letters" and "covenants not to sue" issued by state or federal regulators are important tools, but one at a time protections are time consuming. Environmental insurance has proven to be a useful tool in many projects, providing a level of protection, especially to lenders. Developers need to take advantage of all the protections available, including the site specific as well as the broader protections available on an area-wide, community, state or federal level.
- Institutional controls: Not every molecule of contamination can be removed from many sites. Often contaminants are encapsulated to protect the environment. Such encapsulation and sometimes active treatment systems provide what is called an institutional control, allowing the bulk of the property to be safely redeveloped but incurring ongoing costs and potential risks. While contamination, sometimes in concentrated "hot spots" are removed to approved disposal facilities, *in situ* remediation with institutional controls can also be effective.

In general, environmental regulators appear to be moving towards a more nuanced approach: a clear set of standards for relatively lightly contaminated routine sites, some form of self-certification and increasing flexibility on how to achieve those standards. The New York City Brownfields program, initiated in 2010, exemplifies this approach and has already resulted in fifty redevelopments, some on waterfronts. More difficult and heavily contaminated sites require sophisticated technological approaches such as risk-based analyses and individualized approaches and are most often done at the state and occasionally the federal level. While waterfront brownfields will sometimes require more time-consuming site specific approaches, use of the faster approaches should be considered whenever possible.

F Maximizing the benefits of waterfronts and creating true mixed-use for waterfronts

Many waterfront redevelopments, as reflected in the case studies, are mixed-use. They include a combination of residential, commercial (retail, office and hotel), recreational, maritime and institutional uses that actually support one another. So while each property specific component must be financially feasible, there needs to be synergy, true benefits among these components for the overall project to succeed. This is why expert waterfront design is essential, understanding the interactions between land and water, public access and private use, all the ingredients important for success. Strong consideration of water dependent uses such as marinas are often required by coastal zone plans and may add to the overall value, even if not the highest and best use from an economic value. Stamford's Harbor Point is a true mixed-use development – already having residential and retail uses that support one another, but relocation of a boatyard became contentious. That project also illustrates different market categories are on different cycles; the residential is all rental, no condominiums as initially planned, the office market has been sluggish and the maritime use and public access concerns became controversial, which represents the complexity of benefits and risks of mixed-use development. As in Trenton and other locations, there was a need to re-introduce the community to its waterfront, including special events.

Dealing with a waterfront location complicates remediation as well as design, access and other aspects of the redevelopment. There are approaches that utilize the waterfront. One example is that groundwater/surface water interactions can be exploited. Let water go where it wants to and use features such as tidal/lock and dam influences to be helpful.

Historically, many cities have been built or expanded on to fill, with former wetlands or even open water built up and usually protected by some sort of static seawall. While this "reclaimed land" may seem an expedient solution, places from New York to Hong Kong have effectively banned waterway filling. Wetlands protections, at the federal, state and local level generally restrict wetlands filling or at a minimum require a wetlands mitigation bank – restoring twice as much wetland as will be lost by fill. There are very specific and limited circumstances where filling will be allowed to create or substantially expand a development site.

Waterfronts uniquely allow the use of ferries as a transportation alternative. In cities such as Seattle, ferries are a major transportation element, in other places from New York to Michigan to Vancouver and Florida ferries offer alternative transportation for a relatively small proportion of travelers compared to cars or trains. Among the better known examples are ferries that cross Lake Michigan and Long Island Sound, Nantucket, and water taxis in Jacksonville and Fort Lauderdale, Florida, as well as in the Stamford Harbor Point project. Ferries are far more than scenic; they can service islands (from Shelter Island, New York, to Catalina, California) and be more convenient. Some ferries are primarily or solely for people, while others transport cars and trucks. After the 9/11 attacks and during the San Francisco earthquakes, ferries provided important emergency options. Providing for ferry service can be complex, involving many approvals and substantial capital expenditure, but they are a unique, identifying and often valuable approach to support waterfront development.

Similarly, there are numerous examples of creative design that incorporates flood control. Providing features ranging from a golf course to a sloped shoreline with a secure angle of repose, minimizes hard construction in flood prone areas, while also being an amenity to the project.

Case study: Harbor Point, Stamford, Conneticut

Stamford's South End has historically been a rough mix of industrial and residential uses. As industrial users including Yale & Towne and Pitney-Bowes declined, properties became unused. An early visionary (Arthur Collins), a stalled site (Heyman) and an overly aggressive developer (Antares) all tied to redevelopment properties starting in the 1980s and 1990s and failed. Environmental costs, delays and tightening real estate capital contributed to the problems facing this major urban redevelopment. Then in 2008, with the economy beginning to rebound, strong financial partners (Lubert-Adler and initially Goldman Sachs) who were already involved in the Harbor Point project chose to bring in an experienced developer (Building and Land Technology) to have the project move ahead. Surrounded by Long Island Sound on three sides and on the north with highway and rail service, the 80-acre Harbor Point project plan includes 1,000,000 square feet of commercial space and 4,000 residential units in five distinct areas. The total project value exceeds $3 billion and is one of the largest developments in the Northeast today.

Harbor Point is well underway and has already remade the South End with close to 3,000 units built, a successful Fairway supermarket, banks, restaurants and other retail stores, plus several new or renovated office buildings. The set of "villages" conceived by planner Andy Altman and then of Goldman Sachs are now apparent with most of the buildings designed by Victor Mirontschuk of IBE-International. The project continues to wind its way through a complex set of city, state and federal regulations. The most recent issues have been a requirement to save a local repair boatyard as distinct from a marina, public access and the nature of a proposed hotel. Many aspects of this project are worth noting: a former utility site directly on the water, the nineteenth-century Yale & Town Lock assembly property including historic loft buildings that laid largely vacant for forty years, a master planning process that started with a City plan, and a design for the former industrial facilities of Pitney-Bowes adjoining their I.M. Pei-designed former headquarters. Harbor Point has an extensive open space network, featuring an extensive waterfront walk as well as a central common park.

Stamford is a growing suburban "edge city," forty-five minutes by train from New York. It has reinvented itself from an industrial town to a corporate research community and eventually into a financial center featuring Swiss Bank, RBS, SAC and GE Capital. Most recently, media companies including those with television studios and NBC Universal Sports have grown. The South End enjoys views as far as Manhattan and I-95 access and was long perceived as ripe for urban redevelopment. But the highway along with Amtrak/Metro North rail line separates the site from downtown. There is a major train station served by Amtrak and its Acela is second only to Grand Central in terms of daily trains on the Metro North rail system. It is a five to fifteen minute walk from different residential

Figure 6.2 33 Harbor Point, Commons Square, Stamford, CT
Source: Photograph by Victor Mirontschuk, IBE-International.

areas to the train station, and Harbor Point runs a free bus shuttle to the train station as well as to the downtown area just beyond.

The approval process initially moved relatively swiftly for such a massive project. The Harbor Point team, led by General Counsel John Freeman included highly regarded architects as well as local counsel, planning, housing and environmental experts. Stamford is a sophisticated city and had earlier prepared master plans for the South End, recognizing that the South End was a hodge-podge of outdated and underutilized properties awaiting redevelopment. The project had little initial local opposition, with a relatively small and unsophisticated community. The downtown business community was primarily concerned about competition from the downtown business district. Harbor Point was strongly supported by then Mayor, now Connecticut Governor Dan Malloy. As Robin Stein, the long-time Stamford Planning Director and now Chair of the State Public Utilities Regulatory Authority noted, this developer's approach was to focus on speed and push the envelope in the field. With strong political support and little opposition, many items, such as providing parks, affordable housing, remediation and environmental sustainability requirements were agreed to and negotiated quickly for a project of this size. By moving both remediation and redevelopment simultaneously, the developer was able to achieve a synergy that effectively reduced the cost of the total development. There were issues including: environmental work procedures, certificate of occupancy disputes, efforts to limit waterfront park use to residents only, removal of a full working boatyard, labor practices and the nature of a proposed hotel.

Community concerns, especially the loss of the working boatyard despite clearly stated zoning requirements, which angered abutting boat club members, eventually resulted in the loss of a proposed major hedge fund new office building, but Fairway supermarket was well into operation, and a few other retailers as well as restaurants including a CVS had also opened their doors. The totally rehabilitated 200-unit lofted historic building rented out very quickly. This long, narrow historic building had recently housed only a number of artist studios, which were relocated by BLT to a nearby old, but not historic, structure. Rentals of a new 316-unit high-rise farther from the train station moved somewhat slower. Another 300 units are now being rented. The new, promised Waterside elementary school is open. Two office buildings are complete, but there have been only limited office rentals, despite office being BLT's strongest experience. However, this is understandable given Stamford's now 20 percent office vacancy rate. BLTs South End portfolio also included a number of existing office buildings, one of which has been sold. Harbor Point has achieved a LEED-ND (Neighborhood Development) Gold certification and some individual buildings are also LEED certified.

Perhaps the most interesting financial process was the successful issuance of Tax Increment Financing (TIF) for infrastructure improvements of $145 million, in January 2010. This process was all but complete in late 2009, when Malloy's term was over. The new mayor, Michael Pavia, was cautious and wanted all the bond paperwork (and credit) to be done before he took office. Underwritten by a west coast firm, Stone and Youngberg, the bond includes $16 million under the Federal Recovery Zone program with a subsidized interest rate of 6.78 percent. The remaining $129 million was raised through the sale of tax-exempt special obligation revenue bonds, a class of municipal bonds. Within that category, $113 million of the bonds had a 7.87 percent interest rate and a maturity date of 2039, while $15.9 million had a 7 percent interest rate and a maturity date of 2022. As with most TIFs, the bonds are not backed by the full faith and credit of the City of Stamford, though the City's AA+ rating may have helped the issuance. As a result of the TIF financing, roads, infrastructure and parks have been built this past year. In late 2011, the City received a $10.5 million grant to upgrade the existing train station. It is all about the timing: will the major brownfield urban redevelopment move quickly enough to keep creditors, including the $10 million annual TIF payments that start soon, at bay?

Despite the recent flurry of criticism, Harbor Point has already dramatically changed the South End and has all the elements of urban redevelopment: remediation of contaminated sites, transit access and improvement, historic renovation, new community facilities including a school, innovative financing, new waterfront parks and amenities based on a master plan.

Two case studies, Toledo, Ohio

Toledo, Ohio, is a classic rust belt city, a maker of automobiles and auto parts, including glass, spark plugs and Jeeps. The city, economically tied to nearby Detroit, Michigan, has shrunken though avoided fiscal disaster. Toledo has always been an auto town; car parts from Champion Sparkplugs to Libby–Owens–Ford windshields were made in Toledo. Toledo's most iconic auto manufacturing product was the Jeep, famous for its role in World War II. Toledo-based Willys-Overland played a key role, but ownership passed to Kaiser, then AMC, then Chrysler and now Fiat.

Over the past forty years, the City of Toledo, State of Ohio and the United Auto Workers union have "saved" the Jeep plant in several efforts by giving incentives to keep Jeep manufactured in the city. The latest effort and next decision will be made as the Jeep product line is retooled in 2017. While the current plant on Stickney Avenue and some nearby vacant land is reserved hoping for future Jeep use, there is also considerable land, the so-called South Plant, formerly used by Jeep that is available for redevelopment, located near the center of the city on Jeep Parkway.

The Toledo-Lucas County Port Authority that operates the very active mostly grain shipping facilities on the Maumee River has put forth a redevelopment plan that largely aims to allow expansion of existing industrial facilities, attraction of new businesses and an expansive park along the Ottawa River. Since World War II, Jeep has been at the economic heart of the Toledo community and reuse of this centrally located site will continue to play an important role in the city's future. It was often said that when Detroit, meaning the auto industry, caught a cold, Toledo sneezed. While Toledo has not gone bankrupt, the changes in the auto industry and the latest recession have certainly taken a toll on this classic rust belt city, as population and economic vitality suffered. The old Jeep site was especially challenging as it was heavily contaminated. Both the United States and Ohio Environmental Protection Agencies have contributed technical assistance and regulated the remediation.

The Port Authority, based upon economic planning consultant studies, has initiated a plan for a 111 acre former Jeep site, now called Overland Industrial Park. A joint venture between the Port Authority and locally owned Industrial Developers Ltd was formalized in October 2015. The first building will be a 100,000 square foot industrial facility, utilizing some of the existing infrastructure and significant new road and utility improvements. While the focus is on job creation, the plan that came out of a Community Summit led by Vita Nuova LLC also calls for job training facilities and open space linkages to the community. The Port Authority proceeded with the first speculative industrial building in Overland Industrial Park, in what has been described as a "gutsy call."[31] By August 2016, the Port Authority was able to announce first one, then a second major industrial tenant for this building, bringing hundreds of new jobs to Toledo.

The second project, also led by the Toledo-Lucas County Port Authority using its resources to promote economic development, is in East Toledo, the other side of the Maumee River, where industry including formerly steel mills and still the Port of Toledo and Tony Packo's, made famous by Corporal Klinger of the TV show *MASH*, are located, but where population has been declining.

In Toledo, the east (heavily industrial) side had Main Street and limited waterfront reuse efforts over time. The West (downtown) side of Maumee River saw a series of long-term urban renewal type projects with some successes and many struggles. As Toledo Community Development Director Brad Peebles said about the 2011 effort, "It all started when Mayor Michael Bell went to China in 1998 and came back with potential EB-5 investors to acquire and develop the proposed marina district site in East Toledo." In 2002, First Energy gave around 120 acres of land to the city along with $8 million for environmental cleanup. The site, formerly the home of the Toledo Edison power plant, had many environmental concerns including remediation of fly ash ponds and asbestos removal. Over the last ten years, the city had spent close to $40 million, which included building infrastructure as well as remediation and a new two-foot cover of clean soil. This project has now spanned three different mayors. In 2009, discussions between a Chinese investment company called Dashing Pacific (DP) and then Mayor Bell, former fire chief of the city, began focusing on existing restaurant venues. It led to the $2.1 million purchase of a waterfront restaurant in a former warehouse that was part of earlier renewal efforts. DP then agreed to consider an acquisition of 69 acres of the former First Energy site but shied away after a few city councilmen made questionable public remarks in the news media. The mayor traveled to China to court the company and brought them back to the table. The remediation addressed PCB, heavy metal, coke and emissions. Pellets were designed to promote *in situ* bioremediation but had limited effectiveness in the former lagoons and shallow estuary. Later, large geo-fabric tubes were used to take dredge materials, seal and dewater them to be treated as center of dikes to handle excess water flooding and channeling of the storm water flow that is a problem in the flat, low-lying lake topography. The deal was 60 acres for $3.1 million all cash, EB-5 eligible but no loans. The laws of China state that no one is allowed to transfer more than $250,000 out of the country in a week, so after thirteen weeks, DP had enough cash to buy the land. The land also came with an option to buy another 13 acres, which after remediation by the city, will be turned into green space. Six of those 13 acres have an old Acme power plant on them, which also has fly ash. Their plan is to build multifamily residential while blending commercial space. The model they are using is called the "international village" where stores and restaurants reflecting many different cultures will provide an exciting venue for shopping and recreation. The east Maumee waterfront, long industrial, is being redesigned after extensive environmental assessment and remediation, with new shoreline treatments and extensive upgrades to the Skyway (previously Glass City) Marina. This project has stalled, and it is not clear if it will proceed.

All of these efforts have brought limited change to East Toledo, which remains a poor part of a poor city. The community, long predominately eastern European is now more Hispanic – but still hard-working. The east Main Street retail corridor still exists but struggles. New housing has been limited and most is subsidized; there is some suburban growth to the east and south, sometimes taking good soybean growing farmland.

There are other projects all aimed at bringing new economic activity to Toledo, including the start of the $18 million new warehouse building, part of a 160-acre industrial redevelopment by the Toledo Port Authority. A casino has been proposed about 1.5 miles east of downtown on I-75, but it is not connected to the east side riverfront development. The casino group does not yet have plans to build a hotel on the site but could if room occupancy rates downtown reach around 70 percent. Meanwhile, the Marriott has agreed to put in a Courtyard Hotel in the former Fiberglass tower downtown. The Park Inn, another downtown hotel, has been bought by Chinese investors and is going to be turned into an international business center. There is some long-awaited downtown residential movement happening: 200 market rate single units are coming online, with possibly 250 more coming next year. In other hard-hit parts of the city, urban agriculture is being utilized as an interim use. While there are many moving parts, the east side riverfront is at the core of the city and represents a key component of the rejuvenation efforts. The East Waterfront projects, with extensive environmental and redevelopment costs, will not be possible without the infusion of new capital.

Case study: gas stations

The term urban redevelopment often brings to mind large revitalization projects, but there are also smaller projects that can be significant to a community and to the developer. The redevelopment of gas stations has become a common and important variation of urban and contaminated site redevelopment.

There are close to 100,000 closed gas stations across North America.[32] As automobile and truck traffic became ubiquitous from the 1920s to the 1960s, gas stations proliferated. Not only was there competition among oil companies, but there were locational nuances; left hand turns were to be avoided, some preferred the near corner before the traffic light and others the far corner after the signal; it seems that every busy intersection had several stations. In addition, there were many gas and repair shops along rural roads, famously on Route 66 across the Southwest.

Since the 1970s, the number of gas stations has declined. Changes in vehicles such as: increased efficiency, greater mileage per tank, recently hybrids and electric vehicles all reduced market demand. The business model changed; the "mom and pop" gas and repair shop was replaced by Multi-pump Island, gas-only facilities near highways or became gas and convenience chain stores, such as 7–11. Meanwhile, specialized repair only shops such as Midas proliferated as cars also become more reliable, more computerized and repairs became less frequent but more technical. Zoning and environmental regulations became far more prohibitive since the 1980s; liability and repair obligations resulted in bankruptcies. The result has been closed and often abandoned gas stations in many urban neighborhoods, rural towns and even in suburbia. There are now only about thirty gas stations in Manhattan: risks are too high and profit margins are too low and, most of all, what appraisers call the highest and best use of the property in high-priced real estate markets was no longer a gas station.

Remediation is a major issue for gas station redevelopment, though petroleum contamination is very common and often dealt with quite efficiently. Gas stations have been classified as brownfields and are often eligible for funding and included in community brownfield programs. Before requirements for double-sided, corrosion resistant tanks, there was often petroleum leakage. Other fluids such as transmission or brake fluids and the solvents used to clean tools wound up in the soil and sometimes in the groundwater. While a simple "tank yank" and removal of any shiny petroleum tainted oil may have been enough, now often extensive testing and documentation is required. Oil companies typically have significant liability for tanks and their cleanups, but timing and cost issues often still significantly complicate remediation and reuse.

Redevelopment of former gas stations has become both a business and an art form. While typically done one at a time, there have been portfolio sales of gas stations often related to convenience store operations. Some strong locations become coffee shops, such as Starbucks or Dunkin Donuts. Inner city locations have seen redevelopment as drug chain stores such as Walgreens. Sometimes former gas stations become part of larger commercial, housing or mixed-use projects. Gas stations sometimes occupy important, gateway locations to a neighborhood, and many of these, from swank Bainbridge Island, Washington,[33] to Sanford, Maine, have been reused as parks. Some neighborhood locations across the continent, from Los Angeles to Illinois, have been used as community gardens, though care must be taken that clean soil is brought in and that there is no uptake of contaminants into the vegetables. There are numerous examples of creative reuse of gas stations as art facilities, funky restaurants such as a wine bar in St. Louis and community centers in Youngstown, Ohio, and Winnipeg, Manitoba. There are some iconic gas station designs and a few, such as on Route 66 in Winslow, Arizona, and a Mies Van der Rohe designed station in Montreal, Quebec, have been preserved and used as art centers. One innovative new reuse in Denver is for a marijuana store! The reuse of gas stations is an example of small scale but highly creative reuse and redevelopment.

Notes

1 Superfund or Comprehensive Environmental Response, Compensation, and Liability Act of 1980 (CERCLA) and State laws modeled after this United States Environmental Protection Agency program.
2 Resource Conservation and Recovery Act (RCRA) 42 U.S.C. §6901 et seq.
3 Small Business Liability Relief and Brownfields Revitalization Act (Pub. L. No. 107–118, 115 stat. 2356, "the Brownfields Law") signed by President George W. Bush in 2002.
4 Public Law 107–118 (H.R. 2869) Small Business Liability Relief and Brownfields Revitalization Act, 2002.

5 United States Environmental Protection Agency, David Lloyd, Brownfield Program Directors, December 2011.
6 United States Environmental Protection Agency, ACRES Database.
7 Bartsch, Charlie, Congressional Testimony and US Congressional Budget Office, 2002.
8 HUD Office of Community Planning and Development, Brownfield Economic Development Initiative, 2011.
9 Crocker, Diane and Michael Wallace, EDR (Environmental Data Resources), presentation at National Brownfield Conference, April, 2011.
10 Haninger, Kevin, Lala Ma and Christopher Timmins, "The Value of Brownfield Remediation," National Bureau of Economic Research, *Working Paper No. 20296*, Issued in July 2014.
11 Ibid.
12 United States Government Accounting Office, United States Environmental Protection Agency, ACRES Database.
13 Public Law 96–510 as amended, Comprehensive Environmental Response, Compensation and Liability Act of 1980. Public Law 94–580 as amended, Resource Conservation and Recovery Act of 1976.
14 United States Environmental Protection Agency, Handbook on the Benefits, Costs and Impacts of Land Cleanup and Reuse, 2011.
15 Meyer, Peter, *State Initiatives to Promote Redevelopment of Brownfields and Depressed Urban Areas: An Assessment of Key Features*, United States Department of Housing and Urban Development Office of Policy Development and Research, 1999.
16 Frost, Greta, "Brownfield Impacts on Surrounding Property Values and Overall Economic Health of Cities," by Youngstown State University Student Western Reserve Port Authority, suggests that there can be a 10.45 percent increase in housing prices following remediation of brownfield sites in Mahoning County, Ohio. New Partners for Community Revitalization, "Evaluation of the New York State Brownfields Opportunity Area Program", NYU Wagner Capstone (Tyler Gumpwright, Rose Martinez, Rachel Cohen, Sam Levy, Javier Garciadiego, Prof. Michael Keane), May 2016.
17 www.brownfieldsconference.org/en/Page/156/Phoenix_Awards.
18 www.brownfieldrenewal.com/renewal-awards.html.
19 United States Environmental Protection Agency, Evaluation of National Brownfield Program, June 2014.
20 Hersh, Barry www.brownfieldrenewal.com/story-news-features_perspective_policy_innovation_brownfields_by_the_bunch-921.html.
21 33 U.S.C. §1251 et seq. (1972).
22 www.epa.gov/smartgrowth/brownfields.htm.
23 White House Council on Environmental Quality, February 18, 2010, www.whitehouse.gov/administration/eop/ceq/initatives/nepa.
24 Innocent Purchaser and All Appropriate Inquiry terms defined by federal statute aimed at assuring that brownfield benefits do not go to polluters but rather to independent redevelopers, see www.epa.gov/brownfields/aai/index.htm.
25 NAIOP Brownfield's Tax Incentive.
26 "Brownfields by the Bunch" by Barry Hersh, *Brownfields Renewal*, February 2010.
27 Darden, Thomas, discussion in March, 2009.
28 United States Environmental Protection Agency, Brownfields Office, www.epa.gov/brownfields/aai/aaigg.htm.
29 Northeast-Midwest Institute (E. Paull) The Environmental and Economic Impacts for Brownfields Redevelopment, 2008.
30 www.gao.gov/new.items/d07152.pdf.
31 Linkhorn, Tyrel, Risk with Overland Park Paying Off, *Toledo Blade*, August 28, 2016, quoting Michael Taylor of Vita Nuova, LLC.
32 National Association of Convenience Stores, 2014.
33 Ferry Point, Bainbridge Island, Washington.

7 Revitalizing neighborhoods, housing and social equity

Genevieve Lee Cabanella

Recent decades have seen a dramatic rise in the aspirations of urban communities. As the boomers' suburban generation fades, the millennials have come to seek lively mixed-use transit-oriented urban communities that also reflect more heterogeneous populations. The opportunity for neighborhoods that include a range of incomes, lifestyles, races and religions has grown. Daniel Patrick Moynihan's "melting pot"[1] foreshadowed much of what urban neighborhoods have become, but changes in technology; media and marketing are new, significant factors. The use of mobile devices not only provides interconnections, it makes us safer as connections to friends and police are all but instantaneous. Maximizing social equity in a connected metropolis is both a challenge and an opportunity for urban redevelopment.

The famed urban observer, William H. Whyte, once described cities as "places for the very rich, the very poor and the slightly odd."[2] The poor have always flocked to cities as places that offered more opportunity. While cities do offer great promise, the arriving poor often find limited housing choices, often crowded slums. Finding a place for those with limited resources to live, especially the homeless, has been a challenge for virtually every major city. Racial, religious and social discriminations often exacerbate the issue. The unusual, whether gay, artistic or otherwise different, often find the social fabric of the city relatively welcoming and have even become urban pioneers, finding cheap housing and sometimes creating a market for urban revitalization.

The disparity between the palatial homes of the wealthy and the squalor of often nearby poor results in social and political unrest, and cities continue to be the locale for many widely varied types of complaints and protests. Urban redevelopment is still at times seen as the old urban renewal, removing or damaging existing communities often of color. Leaving urban neighborhoods in poor condition has led to the thought that the only thing worse than gentrification is no gentrification, and some recent studies have suggested that displacement of long-term low-income residents, which is the major concern, may not be as significant as the conventional wisdom suggests.[3] For urban redevelopment to succeed, there must be true community engagement, decision making based upon residents knowledge of what is important to keep and what needs to be replaced. Unlike traditional suburban development, which often aims at one market segment, urban redevelopment almost always aims at providing a range of housing opportunities in terms of type of unit (townhouse, garden apartment, high-rise), ownership (single-family, townhouse, condominium, rental) as well as price, whether rented or owned. Redevelopment may also involve rehabilitation, the improvement of existing housing stock, as well as new construction. Working with a community to provide a range of housing, including for the low income, has become an all but mandatory hallmark of successful modern urban redevelopment, but finding subsidies for affordable housing is challenging. It is also important to note that even a successful redevelopment will not help every person in the community. Programs that focus on individuals, such as for the homeless or for gifted students, and the human part of the equation, are also a key component to real change in a city.

Majora Carter, the founder of the non-profit Sustainable South Bronx and a MacArthur award winner, has created parks, supported better air quality and affordable housing and opposed a jail in her Hunt's Point neighborhood. Served, isolated and environmentally impacted by urban renewal era highways Hunts Point is a low-income, minority community dominated by New York City food distribution hub. Recently, she put forth a concept she calls self-gentrification, focusing on ventures such as Startup Box, a high technology incubator, and now a Birch Coffee shop, Manhattan style but in this underserved Bronx neighborhood, owned and operated by neighborhood residents.[4] A new term, self-gentrification is about redevelopment, creating new employment opportunities, shops, housing and open space, all led by and benefiting community residents.

History of urban renewal and public housing

Public housing was an early, direct approach to providing low-income housing, first authorized in 1937 to combat the Great Depression and later to help veterans returning from World War II. Urban Renewal in the 1940s and 1950s, as mentioned in Chapter 1, was seen as the way to revitalize struggling cities as whites fled to sprawling suburbs; and renewal generally included the provision of public housing. The government's goal was to stimulate the real estate economy through the establishment of various related entities: the Federal Housing Agency, the Federal Home Loan Banks and the Federal National Mortgage Association (Fannie Mae). This was a time during which formal structures for affordable housing were created and developed. One can see the prototypical brick buildings of public housing, following Le Corbusier's Tower in a Park design concept throughout major cities – except where they have been demolished. Over 1.2 million units of public housing now exist throughout the United States. In 1968, the Nixon administration ended the new construction of public housing for families; only a small number of senior housing projects continue to be built under Section 202, and the Section 8 voucher program was a partial replacement for families.[5]

Canada, with its wide open space, has long enjoyed relatively moderate housing costs and much less of a racial aspect to housing needs, with its own very different history of ethnic issues. The Government of Canada, through Canadian Mortgage and Housing Corporation, works with provincial and territorial partners by providing funding to reduce the number of Canadians in need by improving access to affordable, sound and suitable housing under the Investment in Affordable Housing (IAH). Provinces and territories also deliver affordable housing programs that are not funded under the IAH.[6] Canadian cities, including Vancouver as discussed in the urban design chapter, have vigorous and noteworthy urban redevelopment efforts. Toronto is best known for its notable and at times hotly debated urban waterfront redevelopment; a controversial proposal for a waterfront shopping center started the demise of notorious Mayor Rob Ford. While Montreal and Quebec City are perhaps best known for their historic preservation efforts. Without a legacy of nationwide urban renewal, Canadian redevelopments and affordable housing efforts are more varied and led by provincial and municipal governments with significant input from civic organizations.

United States' urban renewal was sometimes characterized as "negro removal" and relocated minorities became predominant in many public housing projects, replacing old

Figure 7.1 Pruitt Igoe implosion
Source: Newman, Oscar, *Defensible Space*, Macmillan, 1972.

tenements with high-rise buildings with the same or worse social problems. Notorious "projects" such as Pruitt-Igoe in St. Louis and Cabrini Green in Chicago became as unsafe as to be uninhabitable and were demolished. As discussed in Chapter 3, architect Oscar Newman was among those who criticized public housing. In 1968, the Nixon administration largely replaced funding for new family public housing with the Section 8 voucher program, which still supports tenants in existing housing units and sometimes in new construction. The smaller HUD Section 202 program continues to provide support for senior citizen housing, sometimes high-rise. HUD stopped supporting high-rise housing for families by the 1970s, and Andrew Cuomo as HUD Secretary for the Clinton Administration created the HOPE VI program that replaced high-rise public housing for families with low-rise, often townhouse designs reflecting both new urbanism and defensible space concepts.

In a notable shift from the blight eradication of urban renewal, today, gentrification, meaning when a neighborhood becomes so successful its own residents are among the less affluent who can no longer afford to live there, is the most significant, critical and intractable issue in urban redevelopment. While old-fashioned urban renewal forced out minorities by eminent domain and other legal tools that are less frequently used in recent years; today's gentrification represents the forcing out of low-income and often minority residents by economic pressures. While there is often a strong, general perception of gentrification in relation to urban redevelopment, particularly in hot, gateway markets, some recent articles have argued that actual displacement of poor minorities by affluent whites is relatively rare.[7] Jerry Brown, three-time California Governor asked when he was Mayor of Oakland, "if we improve policing, make the streets better, and improve schools, rents will go up. Do we not do those things because rents will rise?"[8]

Financing affordable housing

The ingredients that result in successful urban redevelopment – improved transit, brownfield remediation, better open space, creative design, even historic renovations and most of all the attraction of millennials to lively, hip communities – can result in pricing out existing, less affluent residents. The provision of housing for current residents and low-income and minority people in general has become a critical issue. The first response is to try and protect the residents that remain in place, but that requires either limiting rent increases, perhaps balanced to some degree by real estate tax abatements or providing subsidies, both difficult. Providing housing for seniors and families with limited income has become a critical aspect of social equity in urban redevelopment.

Housing subsidy programs are the most often used tools to support provision of affordable housing in today's redevelopment; the most significant federal program today is the LIHTC (Low-Income Housing Tax Credit) program. Created in 1986,[9] administered through HUD and the states and utilized throughout the United States, the LIHTC provides a tax credit for apartment developers who commit generally to preserving rents affordable for residents making less than 80 percent of the Adjusted Median Income in the specific metropolitan area. The LIHTC is somewhat complex; there is what is referred to as the 4 percent and 9 percent types of financial subsidies, associated bond financing and other facets of the program. The LIHTC is now the major subsidy available for both new construction and rehabilitation of housing and has resulted in 2.4 million units nationwide,[10] many within urban redevelopments. Often there are preferences given to residents in the area and for those displaced.

The LIHTC is popular for several reasons, one being that it provides a role for both private developers and community-based non-profits, separately or in partnership, to build and operate affordable housing. LIHTC also allows far greater design flexibility (as compared with public housing) while still meeting HUD standards, resulting in more varied and contextual design. The HUD Section 8 program, which provides tenant subsidies and initially was a partial replacement for new public housing has also been utilized to support existing, rehabilitated and new housing in urban projects. Furthermore, LIHTC encourages public-private partnership in affordable housing development, as the tax credit investor can often partner with the state financing agency in order to maximize debt and equity contributions to the deal.

There is also the Clinton era New Market Tax Credit for commercial projects. This program, while cumbersome, has provided capital for retail and other developments in low-income communities. There are also other tools including Industrial Development Bonds and TIF that have been used to raise funds for projects based upon their future tax payments providing the long-term income stream to repay lenders. Another federal support system for affordable housing aside from LIHTC is the Home Loan Bank, which provides gap funding subsidy. FHLB has eleven national banks and is a member-owned cooperative bank. The overall mission of the FHLB is to encourage community development by assisting and lending to smaller financial institutions, and federal regulations mandate that 10 percent of the bank's annual net profit be allocated to housing subsidy.[11]

Regulations and incentives in urban development

In addition to subsidy programs, another tool to promote social equity and the provision of a percentage of affordable housing is inclusionary zoning. Cities with strong residential markets from San Francisco to Boston have required developers, sometimes in exchange for higher density, tax abatements or preference in the RFP process to provide a percentage of affordable units. Often large, public-private partnership developments include requirements and financial arrangements in support of affordable housing. Often these also can provide what is called affordable or workforce housing, but more rarely assist low or very low-income families.

Numerous communities, notably Montgomery County, Maryland, and other District of Columbia suburbs and several California cities, have adopted their own inclusionary zoning ordinances. Key variations start with the percentage of low-income units, with 80 percent market rate and 20 percent affordable one common formula. Another important provision is whether affordable means, for example, 80 percent of the median income for the region, sometimes called workforce housing which includes young teachers and firefighters in the community, as compared to housing for the very low income, less than 50 percent of the median which would include those receiving public assistance. New York City, under the De Blasio administration, strong advocates of affordable housing, have proposed making a formerly voluntary program mandatory and promoting a 50 percent market, 30 percent affordable and 20 percent low-income model.[12]

Real estate tax policy plays a significant role in land redevelopment, with a set of tools to encourage beneficial redevelopment. Real estate tax abatements have been used to encourage redevelopment, most often as a commercial economic development incentive, but also to promote residential redevelopment. Real estate tax abatements, when used as an economic development incentive, are often tied to job creation, both construction and permanent. In residential redevelopment, tax abatements have frequently been used to support and encourage home ownership.

Tax Increment Financing (TIF) is another real estate tax tool, discussed in the Stamford Harbor Point brownfield case study and in Chapter 8, Real Estate, which provides financial support. TIFs are most often used to provide infrastructure improvement including for suburban expansion, and are based upon the future tax increases from development and are increasingly being used to finance urban redevelopment. As noted in the Brownfield section, many states offer tax abatements and other incentives for remediation that are allowed for (and in a few states give preference to) affordable housing. A few jurisdictions, including Cook County, Illinois, provide tax abatement for contaminated properties which receive state approval of their remediation.[13]

Mixed-use affordable housing

Progressive pursuits in affordable housing concentrate on developing with a mission to resolve social and economic issues in combination with building residential units. Innovative design can be utilized to optimize the effects of healthcare in a more widespread manner and across various fields of public health. For example, affordable housing can be built as part of a mixed-use development together with medical space, which is not only beneficial to residents but to the community as a whole. When these medical spaces encourage primary care visits, this potentially enhances residents' personal health and reduces the economic burden of emergency room visits.

Another option is that the affordable housing space itself can be redesigned to minimize individual residential units and maximize common space, such as kitchens, the lobby, living rooms or lounge space and recreational areas. An example of this would be the Cecil Hotel in Harlem, NY, which is a supportive housing development. Initially, it was converted from hotel use in 1988, more recently was refinanced, and is one of New York City's first ever SRO (single-room occupancy) housing developments that services special needs populations. Transitional and supportive housing can address not only physical needs but also mental and emotional issues, as well as offer individual case treatment. In general, affordable housing can resourcefully redesign space to enable better use of services that could uplift the lives of low- and moderate-income residents, through case management, employment assistance services, healthcare, personal finance counseling and other services. Through the study and observance of results, these socially progressive housing models can eventually be built more frequently and efficiently, thus more appropriately addressing both the improved state of marginalized populations such as the elderly, poor and disabled, as well as overall economic advancement.

Other populations such as teachers and artists may not necessarily be marginalized, but while they may be highly valued in communities, they generally earn only moderate income. Affordable housing can be particularly beneficial to these people but also for the community at large. Housing that targeted to teachers, artists and other types of professionals can be built as part of mixed-use development. For example, a smart mixed-use development would be affordable housing for teachers along with a school building. Teachers Village in Newark, NJ, and Miller's Court in Baltimore, MD, are examples of mixed-use developments with a strong affordable housing component serving teachers. Miller's Court incorporates adaptive reuse with the redevelopment of an historic industrial building that consists of not only affordable housing for teachers, but also substantial space for affordable and collaborate non-profit office space that benefits the surrounding community's needs in education, healthcare and human services. Similarly, Teachers Village in Newark, NJ, is a development consisting of affordable housing for teachers, three charter schools, a daycare and extensive retail space.

Affordable housing is on some occasions also helpful to artists. These include professionals in the arts, such as painters, dancers, sculptors and others. In addition to housing, an ideal urban mixed-use development targeted towards artists would include the component of studios, thereby creating what is called "live/work" apartments. The goal of these developments is to encourage these kinds of residents to remain within communities that value them and hopefully improve local education, culture and overall quality and diversity of the neighborhood's workforce. While targeting housing and mixed-use developments to these special populations is more of the unique case than the common one, public and private investment can be influential in providing programs specific to these populations.

Land trusts, urban agriculture and redevelopment

In the past decade, Land Trusts, sometimes called Land Banks, have become increasingly valuable for urban redevelopment, largely for struggling older industrial cities. Starting in 2004 with the classic auto rust belt city of Flint, Michigan, Land Trusts have now been specifically authorized in several states including Michigan, Ohio, Pennsylvania and New York.[14] These not-for-profit organizations generally acquire vacant urban land, often via municipalities, foreclosed upon for failure to pay taxes. These properties are often blighted and may be acquired, sometimes lot by lot, by Land Trusts, as community-based non-profits.[15] The Philadelphia Land Trust recently took title to its first 150 vacant properties.[16] The Land Trust can clear title, remove deteriorated structures and contamination and sometimes assemble a number of properties to encourage redevelopment.

Some of the policies of Land Trusts go back to Henry George who in the late nineteenth century advocated for higher taxes on vacant real estate, rather than heavy taxation on improvements, thereby encouraging development of vacant parcels. This approach has seen renewed interest, encouraged by, among others, the Lincoln Land Institute of Cambridge, Massachusetts.[17]

Land trusts, often being given properties at little cost and exempt from property tax, can also support beneficial interim use such as urban agriculture, even if the long-term goal is to build housing or community facilities. Urban agriculture, whether by a Land Trust, other community-based non-profit, municipalities or even private parties, is an alternative for the reuse of land with significant health and social benefits. Vacant parcels can be used, on an interim or a permanent basis, to grow vegetables and other plants that provide healthy food and are a positive activity in a community. Urban farming is currently undergoing experimentation in where and how food is grown. Another common example is commercial buildings whose roofs are being used to grow food. Such gardens have long been done by non-profits, often having temporary use of city-owned land awaiting redevelopment. While there needs to be some caution and proper soil technique to insure there is no uptake of harmful contaminants, community gardens can convert a vacant lot into a valuable asset.

Vertical farming is built within the interior, rather than on the open roof, of a skyscraper or high-rise building using the techniques of greenhouse space cultivation. Vertical farming recently has become a popular focus of conceptual design and has been proposed by such famous architects as Rem Koolhaas and Le Corbusier. A real-life vertical farm that was approved in 2015 for development is the company AeroFarms' indoor vertical farm in Newark, NJ. AeroFarms is a large-scale commercial grower for vertical farming and controlled agriculture, and the development will convert a former steel factory for farming use. The farm is a prime example of a variety of methods used to execute urban revitalization: innovative land use, historic adaptive reuse, public-private partnership in financing, and a job creation and training program for the residents of Newark.

Another advancement in urban cultivation is hydroponic farming, which grows plants without the use of soil and instead suspends their roots in nutrient-rich water. This kind of farming is less conventional than the previously discussed methods, but it can be especially used in cities where there are many vacant, unused buildings or where fertile, non-polluted land areas are scarce. Residents of Detroit, for example, are beginning to construct hydroponic farms on empty lots or in old buildings. Other large cities, including Cleveland, OH, and Philadelphia, PA, have vigorous urban farming programs, putting vacant residential parcels back to use.

The total amount of food that can be provided by urban agriculture is somewhat limited and can be very expensive to produce on a commercial basis. The long-term highest and best use of the land, using the real estate terminology, has to also be considered. The Land Bank or Land Trust approach encourages agriculture as an interim use for some properties, but often the midterm or long-term goal is urban redevelopment.

Innovation in urban revitalization

Contemporary urban revitalization utilizes experimental concepts that are built by pioneering developers and result in innovative models that could potentially serve to educate future developers. For example, New Market Tax Credits were previously discussed, and this financing tool is extremely helpful in encouraging new models of commercial urban redevelopment.

Municipalities can also be a major influence on urban innovation, through shaping master planning. Similar to cities and counties that mandate inclusionary zoning, neighborhood planning can also include the requirement of affordable housing. In providing much more detailed specification compared to general inclusionary zoning, master plans often encourage building affordable housing as part of ideal mixed-use urban revitalization, thus enhancing community amenities and achieving urban diversity on a more holistic scale. Communities that utilize innovative design and are put forth as an innovation on district, encouraging entrepreneurship at all levels, can create opportunity. One such master plan was the NYC Lower East Side Seward Park. NYC Economic Development Corporation and NYC Department of Housing and Preservation collaborated on issuing an RFP for the development of this historic and gentrifying downtown neighborhood in Manhattan. The responses were required to specifically include permanently affordable units to families and seniors, consisting of not only rental units but also homeowner units. These would be built as part of a massive redevelopment that would consist of community facilities, significant open space, a rooftop urban farm and notably, creative and tech co-working and incubator space plus a mix of micro-retail spaces. The completion of Seward Park is scheduled to complete in 2024.

Community engagement

Civic engagement, proactive efforts by governments, advocacy groups and real estate developers to communicate effectively with local residents and workers has become a new standard for urban redevelopment. Meetings go beyond not just the legally mandated public hearing, but also design charrettes to long-term efforts to reach a level of agreement, almost never unanimity though sometimes consensus on what urban redevelopment should look like. Urban redevelopment involves engaging the entire community, in terms of design, amenities, housing options, public spaces, retail choices and community facilities.

Meaningful engagement goes well beyond the mandatory public notices and hearings, though those are still relevant. It means bringing the community not just to the proverbial negotiating table but into the redevelopment process. Communication is at the heart of engagement, which today means social and conventional, workshops, educational events, design charrettes and a great deal of personal communication. Elected officials have decision-making authority; their character and commitment to the community are key to meaningful engagement. Public administrators also have important jobs, which includes constant sensitivity and interaction, while moving projects ahead – no mean feat. Some parts of the process, plan creation and approval, the RFP process and implementation are opportunities for meaningful input. Private property owners, business from local stores to corporate outposts, are also important stakeholders. Advocacy organizations; environmental, historic, equality and others are today also stakeholders. Redevelopment projects that succeed have many parents, those that fail are orphans.

The following are two brief descriptions of non-profit community-based organizations that have successfully participated in different ways in various types of redevelopment efforts. There are many such examples around the US and Canada, demonstrating that there needs to be a structure, a level of organization, to help communities actively participate in their redevelopment. These non-profits work with and are generally supported by local government, but they function more directly at the community level.

River Action in Iowa's Quad Cities (Davenport and Bettendorf, Iowa, Rock Isle and Moline/East Moline, Illinois) is a non-profit that has worked for decades to improve the waterfront in these mid-sized Midwestern communities. Projects have included everything from bridge lighting to riverfront trails and bikeways to supporting major new cultural and environmental facilities to flood control issues. River Action, led by long-time Executive Director Kathy Wine, has helped unite four municipalities that are divided by the Mississippi River.

Brooklyn's Fifth Avenue is not like Manhattan's; it runs through mostly working class communities. The non-profit Fifth Avenue Committee has been actively engaged in community improvements, from policing and education, to job training, becoming an active developer, mostly of affordable housing. There is also significant commercial and industrial redevelopment in these communities, where several non-profits are playing a strong role. The former Director of the Fifth Avenue Committee, Brad Landers is now a New York City Council Member and the current Director, Michelle de la Uz sits on the New York City Planning Commission.

The Jacobs Center for Innovation in San Diego, California, was created by a partnership of foundations and community organizations with local government cooperation. The Jacobs Center has a wide range of community goals, including civic engagement, redevelopment, arts and cultural opportunities and economic development. The southeastern portion of San Diego is a multicultural community, including those from numerous Asian and Hispanic backgrounds as well as Black and Caucasian residents. The Jacobs Center provides a range of services including the Diamond Educational Excellence Program and Full STE[+a]M Ahead aimed at promoting a creative community. The Center has also supported the redevelopment of the 45-acre formerly blighted Village at Market Creek that provides new affordable housing, revitalizes a retail center featuring local and healthy foods, and the restoration of 3,000 linear feet of wetlands. The Jacobs Center received an EPA Brownfields area-wide planning grant, one of twenty-three pilot projects initiated by Mathy Stanislaus as EPA Assistant Administrator for the Office of Land and Emergency Management (formerly Office of Solid Waste and Emergency Response) and was cited in recent EPA publications.[18] An interesting aspect is that the foundation support is scheduled to end in 2030, at which time the organization will need to become self-sufficient.

A new tool that combines both progressive community-based lending as well as potential civic engagement is real estate crowdfunding. Today, this enables mostly smaller real estate developers to bypass traditional providers of both debt and equity, through the use of modern software technology and social media. Developers and investors are connected via web crowdfunding platforms, and while the funding platform still underwrites projects and vets investors, the process is much more open and direct than conventional real estate finance sources. Local investors can more directly acquire partial ownership or mortgage interest in a real estate development. Therefore, it is possible to acquire funding or investment from a multitude of owners who have an interest in enhancing their own community's development. The benefit to investors, aside from net gains, is the ability to directly affect and have a stake in development within their communities.

Case study: Camden, New Jersey

Camden, New Jersey, had the dubious distinction of being named the poorest city in the United States, certainly poorest in New Jersey.[19] Today, Camden is moving forward with new developments mostly near downtown, including investments by major real estate firms including a proposed $1 billion campus by regionally based Liberty Property Trust and the redevelopment of the former RCA Victor building by Dranoff Properties. Camden still has

Figure 7.2 Streetscape, Camden, NJ
Source: Photograph by John P. Sullivan.

many problems and it is relevant to consider how this happened and how long it took for this level of progress.

Located on the east side of the Delaware River, near Philadelphia and originally home to Campbell's Soup and RCA Victor, Camden's economic base largely fled; its more affluent residents moved to nearby suburbs, and what was left were the Camden County offices and courts, the port and undesirable land uses such as scrap yards. Three of Camden's mayors have been jailed for corruption, the most recent being Milton Milan in 2000. From 2005 to 2012, the school system and police department were operated by the State of New Jersey. There have been redevelopment efforts led by a local development corporation and a local church, notably the New Jersey Aquarium and related downtown improvements.

In 2004, Cherokee Investment Partners, then the largest brownfield redeveloper in the United States, envisioned a sweeping $1.2 billion redevelopment along Cramer Hill neighborhood in Camden and adjoining Pennsauken. It was in some ways an inspired design vision: a new waterfront community designed on sustainability principles and promises to benefit existing residents with jobs and new homes. There were problems from the start: the use of eminent domain to relocate 700 homes in what was among the more stable communities in Camden, the promise of $175 million of state funding, ecological issues with a nesting bald eagle on Petty's Island and more. By 2006, as the real estate bubble was becoming visible, there was community disenchantment, some race related, legal missteps and the project had stalled.

Over decades, Camden has seen the whole gamut of subsidized housing projects. The City of Camden has a Public Housing Authority has a dozen different communities, ranging from high-rise for senior citizens (including an assisted living program) to family housing in a range of building types. The Camden Public Housing Authority also administers an active Section 8 program. Some Camden public housing has been modernized utilizing HUD funding, including substantial rehabilitation under the HOPE VI program. There have also been LIHTC projects such as the 32nd Street development.

In Camden, single-family homes are still the most common form of housing, and a significant amount of assistance has been focused on improving houses throughout the city, including the Camden Hill neighborhood. These programs have included weatherization, owner advising to avoid or deal with foreclosure especially since the financial crisis. Camden also has private non-profit housing agencies including Neighborhood Housing Services, Catholic Charities and individual churches. A church led group led and put forth the South Camden Community Plan and there are new efforts in the Parkside and Fairview as well as Cramer Hill neighborhoods.[20]

In the past few years, there has been a set of specific, perhaps less grand but important development projects; a process with greater community engagement through neighborhood and church based groups that have had some positive results. Camden, with significant support from the State of New Jersey, and use of many state housing, environmental and economic development programs, has continued to make step by step progress in Cramer Hill, Parkside and other neighborhoods.[21] The Camden Sewer and Water Authority has supported infrastructure and brownfield improvements. Coopers Ferry Partnership, starting downtown with the New Jersey State Aquarium across the Delaware River from Philadelphia, has completed a series of successful projects, in conjunction with the City's urban redevelopment agencies. Projects include a major greenway, creation of a park on a former landfill in the Cramer's Hill neighborhood and the more recent projects culminating in the major announcement of the downtown redevelopment project by Liberty Property Trust. There continues to be improvements, especially in the downtown waterfront, brownfields were mapped, EPA assessment grants received and some cleanups. The County Sewer and Water facilities, located on the waterfront, helped in improvement and remediation efforts.

In July 2015, the seventy-five unit affordable housing development, River Village in Pennsauken, was completed by Conifer, an affordable housing specialist. This was enough of an event that not just local and state officials, but U.S. Senator Cory Booker attended the Grand Opening. Major projects have been announced near the successful state aquarium. Camden and Pennsauken remains a cautionary tale – big visions, even those that are well-intended and well-capitalized, need community support. Incremental steps may seem to take longer but eventually do achieve progress.

Notes

1 Moynihan, Daniel and Nathan Glazer, *Beyond the Melting Pot*, Harvard University Press, 1963.
2 William H. Whyte, *Building the Nation: Americans Write About Their Architecture, Their Cities, and Their Landscapes,* edited by Steven Conn, Max Page, 2003.
3 Ellen, Ingrid Gould and Katherine M. O'Regan, "How Low Income Neighborhoods Change: Entry, Exit and Enhancement," New York University Furman Center for Housing and Real Estate, September 1, 2010. US Census Bureau Center for Economic Studies Paper No. CES-WP-10–19.
4 Sheir, Rebecca, "Self-Gentrifying in the South Bronx," *Slate*, June, 2016 plus Conversations with Majora Carter and Barry Hersh.
5 HUD programs, 2015.
6 www.cmhc-schl.gc.ca/en/inpr/afhoce/fuafho/iah/afhopracca/.
7 www.slate.com/articles/news_and_politics/politics/2015/01/the gentrification_myth_it_s_rare_and_not_as_bad_for_the_poor_as_people.html.
8 Jerry Brown, Waterfront Center Conference, Oakland, California.
9 Tax Reform Act of 1986 (TRA) (Pub.L. 99–514, 100 Stat. 2085, enacted October 22, 1986).
10 Office of United States Controller of the Currency, Community Development Highlights, 2014.
11 Federal Home Loan Bank Board, 2015.
12 Been, Vicki, New York City Commissioner of Housing, New York City Housing Plan, 2014.
13 Hersh, Barry, "Real Estate Tax Policies and Brownfield Redevelopment," Lincoln Institute of Land Policy, Cambridge, MA, 2002.
14 www.thelandbank.org/history.asp.
15 Center for Community Progress, Itelander, December, 2015.
16 www.sfgate.com/news/article/Philadelphia-Land-Bank-gets-first-150-vacant-6687221.php.
17 Hickey, Robert, "The Role of Community Land Trusts in Fostering Equitable, Transit-Oriented Development" (Working Paper) Case Studies from Atlanta, Denver and the Twin Cities, June 2013.
18 Brownfield Area-Wide Planning Pilots, Ideas and Lessons Learned for Communities, 2014.
19 http://newsone.com/3115808/camden-nj-america-poorest-city-obama-police-initiative/.
20 Camden Works, 2006.
21 Camden Works, Camden Department of Development and Planning, 2011.

8 Real estate and capital markets

Rick Mandell

Money, whether debt or equity, is the cornerstone of real estate – and the goal is to make more money than it costs. Violent economic and political cycles, unanticipated costs and unforeseen circumstances are some of the determinants of whether this is possible. The growth of urbanization, worldwide and in the United States, has strongly impacted real estate markets and opportunity through 2007. Starting before but growing stronger coming out of the recession, there has been capital available for urban redevelopment from traditional as well as newer sources such as equity and hedge funds, global financing and crowdfunding.

Many invest in real estate programmatically. Along the risk spectrum, there are institutional investors like REITs (Real Estate Investment Trusts, now a $3 trillion capitalization), insurance companies, pension funds and their advisors, and private equity funds and others who are more or sometimes less, expert at creating and managing the cash flows from real estate. If there are stable markets behind the cash flows, risks, including smaller returns and even failure to repay principal, can be minimized. REITs and equity funds, including Arcadia REIT and Thor Equities now specialize in urban redevelopments and Centerpoint Properties Trust, which have successfully redeveloped former arsenals and other brownfield properties.

There are creators of real estate value who buy land, entitle it, develop it and harvest the value. Development is a high risk, and sometimes high reward, business. Urban redevelopment, the reuse of property in cities and inner suburbs, is most often seen as an even higher risk. If real property, especially in urban areas, is managed poorly, or if shocks occur at the property or financing levels, there are other investors who recognize those events and see an "opportunity," albeit riskier, to fix the issues and harvest higher returns.

The evolution of the influence of disciplined investing in real estate is clear. From secondary markets to "friends and family" country club pools, results are tangible, perhaps sometimes predictable, but nevertheless measurable. The latest real estate phenomena, crowdfunding is the internet-enabled expansion of the same type of real estate investment.

The USA has grown of age with the baby boomers and their progeny, with the end of World War II, two million vets helped produce a bulging economy requiring the development of a specialized subset of entrepreneurs who learned during war effort.[1] Over the succeeding generations, perhaps as the number, types and size of buildings grew along with the opportunity to understand and appreciate risk, large pools of capital were formed to take advantage of these opportunities and create value. First, banks, insurance companies and wealthy individuals, and later pension plans and endowments, eager to meet cash needs and make outsized returns, entered the real estate capital markets. More recently, REITs, TICs (Tenancy In Common, a form of real estate ownership and investment), Pension Fund

Advisors, Private Equity Firms. Opportunity Funds, EB-5 Visa financing (recently extended through April, 2017) and most recently crowdfunding mechanisms have all jumped in to solve capital needs as the real estate gets developed. As the newest form of real estate financing, crowdfunding may be particularly suited to urban redevelopment, relatively small projects that may be attractive to a specific group of investors. Revitalization News recently reported that the town of Valdese, North Carolina, and the Historic Valdese Foundation are attempting to use crowdfunding to restore the Meytre Grist Mill at McGalliard Falls Park.[2]

Whether from extraordinary profits from picking through the ashes of the RTC (Resolution Trust Company 1989–1995, which disposed of $394 billion of distressed real estate assets), the Great Recession (2007–2010) or the coming uncertainty, seeking safety and return is the first priority of the money. In order to access the money, one has to want to understand what the money wants.

In the more recent recession, the heaviest foreclosure rates were in suburban, sunbelt single-family home markets. In many urban areas there was less volatility, but still foreclosures where the economy was weak, cities such as in Akron, Flint and St. Louis. Some commercial and industry properties were also impacted. Foreclosed properties in urban areas did offer both development opportunities and challenges, and there have been a wide range of projects, from new private development in strong locations, to major public-private developments where government support was available such as Slavic Village in Cleveland,[3] to all but abandoned urban neighborhoods where urban agriculture has taken hold, including Detroit, which is reported to have a quarter of the tax parcels in the city 84,000[4] vacant as is described by Rod Stevens in Chapter 1.

In addition to REITs, mortgage-backed securities are another way for real estate to access public capital markets. The Commercial Mortgage-Backed Securities (CMBS) market boomed from 2000 to the 2007 financial crisis and is now re-establishing itself for commercial real estate including rental apartments. Residential Mortgage-Backed Securities, always led by Fannie Mae and Freddie Mac, were far more devastated by the financial crisis, and now these Government Sponsored Entities have an even larger share of the residential mortgage market. The mortgage backed securities market has always preferred uniform products, so urban redevelopment efforts often do not meet their underwriting preferences.

Experience representing real estate private equity funds for over a dozen years indicates that along the risk spectrum, there are many ways to make money. For those seeking safety only, perhaps one measure is the difference between an investment in government bonds and any other investment. In the seven years since the Great Recession began that has not been a very high bar since government returns have been less than 1 percent. So, for example, receiving a 4 percent net return for an investment in a class A office building in a class A location with long-term leases signed by high credit long-term tenants might be considered good, thus many institutional investors were satisfied with such returns.

A step to the next bar of creating value might be that same building with soon expiring leases where the next tenant might require costly tenant improvements, or for tenants which don't require as much space, necessitating expanded leasing commissions and tenant improvements, which will depreciate the investor returns over time.

Perhaps at the riskiest part of the value creation spectrum now is raw land, the heart of traditional suburban development. Valuing a current investment in land is tricky. It involves calculating how to recognize appreciating risk turns on asymmetric knowledge, information everybody else overlooks. Land has no value until it becomes useful. In fact, it has depreciating value – taxes and insurance. Political risk, development risk, market timing risk in bringing it to market, construction risk and lease up risk all contribute to the specialized knowledge that investors must rely on to determine what value the land holds for a current investment. By comparison, urban redevelopment projects, often dealing with emerging established locations and tenants, can offer high potential returns but requiring specialized risk management in terms of product and marketing.

How then investors think about investing in this part of the real estate risk spectrum, especially formerly occupied land? Let's suppose that a house will be built on the land that is affordable to a person earning enough to qualify for a home costing $500,000. Before the house is sold to the buyer, the land has to be zoned, which means that it needs a political decision from an elected and/or appointed body politic – the planning and zoning commission and/or the city council/county commission. If the land is part of a larger parcel not zoned for housing, the zoning may need to be changed, usually a long, risky, and cumbersome process. It might also need to be part of a planned unit development which needs approvals from the state and/or federal governments for environmental issues (pollution, endangered species, wetlands, etc.). This process takes a lot of time, perhaps years or in some instances longer in urban redevelopments.

Once approved by the government, the site needs to be developed with roads and utilities, more consecutive time before a return is realized. Urban sites often offer advantages as infrastructure as well as zoning, may be in place and usable. If it is part of a large development, then it needs amenities to be developed at the same time, another cost. Swimming pools, golf courses and clubhouses are an additional burden on cost for suburban development, in urban settings promenades along a waterfront, transit facilities and community amenities become part of the negotiated development costs. Going to vertical construction takes more time, perhaps as much time as the rest of the development process, if it involves multiple structures built in phases and/or big buildings.

Then, of course, when the house is ready to sell, is there a market for the product at that specific moment in time? Or will interest on the unsold house use up any profit possible? Up until 2006, more or less, with the exception of regular recessions, there has always been a market for most things that were built. Some had to wait and change hands more than once, each at lower going in values. Perhaps the best news in terms of urban redevelopment is that the growing millennial market has a preference for urban, gritty, historic, sustainable places.

Real estate provides investors with the opportunity to obtain measurable risk adjusted returns. How risk is measured is the trickiest part. Perhaps the easiest measure to understand is to hear from the money about what the money wants. There have been numerous funds set up by astute investors such as Goldman Sachs and Jonathan Rose who have been investing in city redevelopment.

As the moderator of the panel discussion for the last five years at the National Association of Homebuilders, I've been tasked with creating the agenda that brings the money to talk about what they want. Private equity, bank debt, understanding what a spreadsheet means and a chance to meet the money and shake hands for future business is the result, though crowdfunding is a new, online variation.

Understanding that the equity that invests will be comparing any investment with all opportunities it sees and will measure that risk/return as an unlevered yield on all in cost is take the first loss position. Explaining that concept to those who actually develop and build, especially in urban areas, is an interesting process.[5]

Funding the gap

A higher proportion of urban redevelopment projects compared to traditional real estate investments require some measure of government support, though this is changing in many strong markets. While land use and environmental regulations apply to all, urban redevelopments often do not "pencil out" – are not sufficiently profitable as projects by a real estate discounted cash flow pro forma. Some form of government support is necessary to fund this gap – but governments also have significant incentive to support urban redevelopments for political, socio-economic and long-term economic development advantages.

The most traditional form of assistance for redevelopment is for the government to bear infrastructure and land costs. Urban renewal used eminent domain to acquire land, write-down the cost, make infrastructure improvements and then sell for vertical development. This approach is far less likely to be utilized in either urban America or Canada today. There are still times when government-owned land or rights, such as railyards or unused municipal property (not parks), property taken for failure to pay real estate taxes, or old urban renewal sites are made available for redevelopment.

Today, it is far more likely that an urban redevelopment will be a public-private partnership, a contractual agreement that covers not just land use and environmental impact, but amenities including transit facilities open space, cultural amenities and the entire redevelopment process, including financial guarantees. The government may provide important improvements, such a light rail, area-wide remediation or resiliency improvements, but the developer builds the building. Often these public-private partnerships include financial agreements, specifying who will pay for what and when.

In some cases, the government may use a program such as Industrial Development Bonds or Tax Increment Financing to help provide what are in effect loan financing for aspects of the redevelopment. Industrial Development Bonds are effectively low-cost mortgages for job creating projects, issued but not guaranteed by the local or state government and exempt from income tax, therefore low cost. TIF essentially borrows against the future real estate increases due to redevelopment and is not considered a debt of the municipality.

Some urban neighborhoods have major institutions, usually hospitals, universities or both, that can play a key role in redevelopment; examples include the University of Pennsylvania in west Philadelphia, Johns Hopkins University and Hospital in Baltimore and the University of Illinois at Chicago.[6] These institutions today often play a key role in some urban redevelopments and can be building blocks for revitalization – but not obliterate its community. Anchor institutions can be partners in development, provide tenants and amenities and financial stability to redevelopment projects. An example of a successful redevelopment with a different type of institutional source is City Creek Center in Salt Lake City. The project is an undertaking by Property Reserve, Inc., which is the commercial real estate division of the Corporation of the President of The Church of Jesus Christ of Latter-day Saints (LDS Church) and Taubman Centers, Inc. Like many other center city redevelopments, there is a combination of retail and residential use, with foliage-lined walkways and streams covering three blocks in downtown Salt Lake.

Some of the financial tools available as part of a public-private agreement that help fund the gap include the use of specific programs mentioned in earlier sections. Historic tax credits may be used for the preservation of qualified structures within the project. LIHTCs are used to provide affordable housing. State and federal funding or tax credits for remediation of contaminated properties. Other government funds, including those focused on infrastructure, may also support urban redevelopment. While shrinking in recent years, there are still numerous other specific programs, for example, health facilities may assist an urban redevelopment.

Real estate marketing

In general, real estate debt and equity investors still look at urban redevelopment a bit skeptically, which in real estate finance terms means a higher capitalization rate signifying greater risk than the most traditional real estate products. There are a set of risks that an urban redevelopment must address in order to obtain financing at rates low enough to allow the project to move forward.

- Safety and security: Redevelopment projects often market themselves places to "live, work and play"[7] – all best done where people feel safe and secure. Urban redevelopment often occurs in places with reputations for crime and a lack of security. To succeed, redevelopment must change both the reality and the image. The approach to providing security used in many projects was first articulated by architect Oscar Newman,[8] providing a high level activity, eyes of the street, visibility and design that allow residents and merchants to manage their own community. Of course, proper policing and enforcement is essential.
- Education: To some degree, urban neighborhoods, as compared to suburban developments, often appeal to those without school-age children, but reaching a broad market of those who do or might in the future have children is important. Some large projects incorporate new schools, often elementary and sometimes charter schools, to improve local education.
- Stigma: Some locations are viewed as less desirable due to real or perceived contamination. Real estate valuation studies have shown that areas with known environmental contamination are very likely to have reduced property value,[9] though it has also been shown that such stigma diminishes over time. Often values will rebound after a single occurrence, but repeated events and publicity take much longer to recover. Also, proper remediation, including receipt of "no further action" letters or other forms of regulatory approval, as well as environmental insurance, can go a long way in convincing potential users and their attorneys, as well as local planners, communities and lenders, that the environmental contamination issues have been properly dealt with and are no longer an issue. The EPA Superfund reuse program is testimony that even heavily contaminated property can be cleaned up and reused.

So while there are challenges to overcome, there are also advantages that urban redevelopments have in terms of marketing the eventual development. Redevelopment, designed properly can have an authenticity, funkiness if you will, that no brand new development can offer. Rehabilitating historic structures, contextual design and local historic references can all offer unique strengths. Being in a real place, with amenities, character and convenience means something to young people. The core advantages of redevelopment often include location and transportation access. These are very strong marketing factors, especially with millennials. Being able to walk or have convenient transit access to employment is a significant factor.

Developers' perspective

Most real estate developers are not ideologues; they are interested in making a profit. Some see value in sustainability and resilience, many appreciate the benefits of energy performance and others are committed to their home markets. Developers look for opportunities that have a strong return and where they can effectively manage risk. For decades, the suburbs represented development opportunity for home builders, office and retail developers; growth occurred on the urban fringe and sometimes in what Joel Garreau called "edge cities"[10] such as Falls Church, Virginia, and Morristown, New Jersey. Urban redevelopment is occurring in many older suburbs, such as Harrison, New Jersey, and Bethesda, Maryland, both of which are being revitalized by major TODs.

Developers would like the modern equivalent of Alexander's mythical sword to cut through the "Gordian Knot" of waterfront brownfield approvals. Case studies and research do not suggest that any one approach, power broker or legal process can magically make a project happen. There are a set of techniques to help more effectively manage the waterfront brownfield set of complexities.

Since 2000, and especially after recession triggered by the residential mortgage collapse, developers are finding more opportunities and less risk. Toll Brothers, the large, public traded home builder primarily built large, fairly expensive houses, their motto "America's Luxury Home Builder;" battling in Princeton, New Jersey, and elsewhere to get subdivisions approved. Now much of their business, called Toll City Living and Apartment Living is high-rise, still fairly expensive, condominium and apartment developments in Brooklyn and other strong urban markets. Advance Real Estate, based in suburban New Jersey, was through the 1980s a successful developer of speculative (no anchor tenant) suburban office buildings. Now their projects include urban infill residential development in Hoboken, a major mixed-use TOD at the Harrison 'New Jersey' train station, next to Newark, and the mixed redevelopment of an old shopping center. When asked about the change, Peter Cocoziello, the founder and head of Advance, responded that he "did not lose sleep"[11] rather he followed the market looking for less risk and more reward. Post Properties in Atlanta was another substantial home builder who has chosen to focus on townhouse and apartment projects that are more urban infill than exurban sprawl.

Economic development

Most communities and jurisdictions, city, suburb or rural, rich or poor, selectively seek economic development. Most often the main goals are high quality jobs and additional tax revenues (real estate and sales taxes), as well as other goals in terms of quality of life and constraints in terms of environmental impact and reluctance to change.

Urban redevelopment is often a key strategy of economic development, seeking to reuse land. The classic example is an industrial town seeking to replace what had been a major employer that closed. While replicating the former use may not be realistic, using a property that is well located, especially in terms of infrastructure, may be an important opportunity. Finding new uses that generate jobs and tax revenues are an intrinsic function of both economic development and urban redevelopment. The types of urban economic activity and the opportunities provided can vary widely, as can the tools used to attract and generate these activities.

Most desirable are core economic generators, activities that an urban region exports and are the basis of a successful metropolitan area. Economic developers, such as Michael

Porter[12] speak of clusters, related businesses and institutions that multiply economic impacts. A classic industrial cluster would involve a manufacturer that needs various suppliers that go into the final product. Today's economic clusters often involve service providers and include institutions as well as businesses. Silicon Valley in California famously began with the creative brain power largely out of Stanford University starting what became major technology companies, initially manufacturers such as Hewlett Packard, but later featuring companies such as Google, Electronic Arts and Facebook, as including all the financiers, software developers and digital marketers that have grown exponentially. That growth has resulted in redevelopment throughout the region, including a host of urban redevelopments in San Francisco, Oakland as well as smaller urban communities in the Bay Area region. Silicon Valley exports technology goods and services to the world, supporting high paying jobs and tax revenues with limited negative environmental impacts.

Urban redevelopers today often focus their efforts on creating high tech and high paying jobs, using concepts such as Innovation or High Tech Districts, a concept supported by the Brookings Institution among others.[13] Sometimes affiliated with research universities, these type of efforts have sprung up in many cities, from St. Louis and Brooklyn to Wilkes-Barre, where a consortium of smaller institutions are supporting regional growth through technology. The concepts that place matters, and how each community must define its strengths, cultivate its economic clusters and that growth is largely in new technologies and job categories, have grown in importance in the economic development community.

Urban redevelopment examples include Cleveland's Health and Technology Corridor, which includes major hospitals and universities and has supported related businesses as well as institutions. Pittsburgh reinvented itself from "Steel City," retaining large corporations such as Alcoa and PPG, and expanding health and high technology clusters utilizing their base of outstanding universities.

Many urban redevelopments involve non-core activities, services such as retail activities. Home Depot saw the opportunity for reusing brownfield sites for their stores, which are big boxes on concrete slabs with substantial parking and provides hardware to contractors and consumers. One example is the Home Depot right off I-95 developed on a former factory site in Fairfield, Connecticut, where it would have been hard to find and gain approval for a comparable "greenfield" location.

Finding the right economic uses for an urban redevelopment can become highly nuanced. Will local residents be able to obtain high tech or health care jobs? Will the pay of new jobs, such as retail, be sufficient, a so-called living wage related to the movement towards the $15 per hour minimum wage? Will boutiques and expensive restaurants gentrify the area? Are fast-food and dollar stores a community improvement? All this ties in to both housing and transportation concerns. Some see the retention of industrial jobs, accessible to those with limited education, as a key element in urban redevelopment.

Urban economic development provides key tools for specific urban redevelopment projects. One key tool is how economic developers analyze growth opportunities, often thinking about economic clusters, businesses and institutions that support one another and result in economic growth. Economic development tools are critical to individual urban redevelopment opportunities. Many states, counties and municipalities across the United States offer Industrial Revenue Bonds (IRBs), usually via an Industrial Development Agency, there are also many state and local Community Development Finance Agencies that offer financing vehicles for infrastructure as well as economic development. Qualifying bonds are considered municipal bonds exempt from federal income tax, and often also state income tax. These are revenue bonds, based and providing recourse from the project, and are not general obligation bonds of the issuing government. Often the underlying property is officially transferred to the Industrial Development Agency and there is an agreed upon Payment in Lieu of Taxes (PILOT) to the school district lower than normal school district real estate taxes, providing another advantage. When the bonds are paid off, the PILOT has expired and the property reverts to the private owner.

Another set of frequently used popular economic development financing tools are based upon real estate tax policy. Many municipalities offer reduced real estate taxes to encourage redevelopment – frequently used in urban redevelopment. Usually these tax abatements are for a specific period of time such as ten years, in some places the tax abatement decreases over that time, so for example a 50 percent tax reduction the first year, 45 percent the second year, so the tax abatement "fades away" and is over in ten years.

Another economic development tool is Tax Increment Financing. First created in the Southwest to help promote suburban development, it has been largely adopted and often adapted for urban areas. An area needing redevelopment, not completely unlike old urban renewal designations, is designated. The state authorized municipality issues bonds that are to be paid back from the increased taxes to come from increased development. The municipality continues to receive the current base taxes (which may have declined in a blighted district) until the bonds have expired, at which time, often ten years or more, the municipality receives the full taxes, as illustrated.

TIF bonds are considered municipal bonds, like IRBs, and not subject to federal and often state income taxes. Also similarly, TIF bonds are not general obligations of the municipality, recourse in the event of default is against the properties in the district. TIF has become a very popular tool (as in Chapter 6, Stamford Harbor Point case study), supporting redevelopment of areas with vacant, abandoned, brownfield and other troubled properties, often in low-income communities. TIFs have been utilized to provide relatively

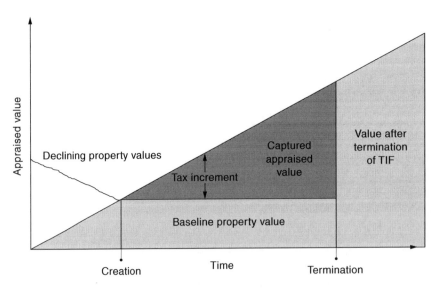

Figure 8.1 Tax Increment Financing illustration
Source: Author.

low-cost financing for urban redevelopments across the United States, including Chicago, Illinois, Albuquerque, New Mexico and many others. California was the first state to authorize TIF in 1945 and first to utilize in 1952, but in 2012 ended redevelopment agencies that had used TIFs extensively, due to state and municipal budget issues. A 2014 state law was aimed at reinvigorating the use of TIF in a revised form.[14] Some provinces in Canada such as Manitoba also allow a form of TIF.

Business improvement districts

Another tool for economic development is the use of Business Improvement Districts (BIDs). Unlike TIF districts and real estate tax abatement programs, BIDs are generally created by a vote of property owners, agreeing to pay limited *higher* real estate taxes that are earmarked for services and improvements within the district, and generally each BID has a governing board largely of property owners and including municipal representatives. BIDs are most often commercial areas, where the owners and tenants benefit from additional marketing, street cleanups, lighting, special events, sign improvements and sometimes safety patrols. The first BID was in Toronto, Canada, in 1970, and first in the United States was in New Orleans in 1974.[15] There are now over 80 BIDs in Toronto, 67 in New York City and over 1,200 in the United States and Canada.

BIDs have proven to be a valuable tool in urban redevelopment. Early and well known BID successes include the Grand Central Partnership initially led by Dan Biederman using many of the ideas put forth by William H. Whyte.[16] A closely related non-profit led restoration of Bryant Park, from an unsafe, underutilized space to a vibrant, activity-filled, extraordinary urban open space. Along with BID improvements ranging from information kiosks to shared newspaper distribution racks, the overall result has included millions of square feet of new office buildings, who proudly use Bryant Park in their name and support the BID.

A recent study of fifteen BIDs in ten states measure BID characteristics and performance, and found successes that spurred redevelopment from the Yerba Buena Community Benefit District in San Francisco to the Ballston BID and Clarendon Alliance in Arlington, Virginia.[17]

BIDs encourage owners and their tenants to upgrade both individual properties and a usually commercial district as a whole. Among the programs frequently provided by BIDs are marketing for the district, sign improvement and coordination, more frequent street cleaning, sidewalk and landscape improvement, improved, lighting and other security improvements. BIDs frequently sponsor events, such as street fairs and shopping days to support businesses in the district. Public spaces, such as parks, bike lanes and promenades are also often improved by BIDs.

Redevelopment real estate taxes and liens

Real estate tax policies significantly impact urban redevelopment in many ways. Often older, urban communities have higher tax rates and sometimes a declining tax base. The high rate discourages redevelopment, while the declining base makes redevelopment crucial to the financial future of the community. As a result, there are numerous ways that local governments try to use real estate tax policy to promote desired redevelopment.

The most straightforward approach is tax abatement, essentially rewarding redevelopment rather than penalizing new development with higher taxes. There are various formats for such abatement, sometimes offered as part of designated redevelopment or empowerment zones, such as offered in Cleveland, Ohio, and other cities. A few cities, including Cook County's Chicago, provide real estate tax abatement for an approved remediation of a brownfield.[18] Another approach is to allow abatements for developments meeting specific requirements such a site not utilized for a set number of years or a development that will create a specific number of jobs. Tax abatements normally are for a specific time period, sometimes declining over time, so the property will pay full taxes at the end of the abatement period. TIF is another real estate policy that can be used to promote development, as discussed earlier, but it should be noted that a city can offer tax abatement or TIF – but not both.

Often projects that are financed by IRBs aimed at supporting projects that produce new jobs are allowed to make a PILOT. As the bond issuer, the Industrial Development Agency, is a public agency and actually holds title to the property which is leased back to the user, normal real estate taxes do not apply. Instead, the project user makes a specified PILOT payment. PILOTs affect not only municipal taxes but also school district tax receipts, so they can be controversial and are legally spelled out in detail. One common format is for the IRB project to pay 50 percent of normal real estate taxes the first year of operation and then an additional 5 percent each year and is making full tax payments after ten years.

A related issue is the transfer of vacant properties which have not paid real estate taxes. An unpaid tax bill becomes a tax lien after a specified time, and local government can foreclose using that lien. Traditionally, cities would foreclose properties for failure to pay taxes, there would be a mandatory redemption period where the owner had a last chance to pay off the taxes, and then the city would auction the property off on the courthouse steps. In recent years, many cities have gone to a different, quicker approach, selling their tax liens, often in packages through investment banking firms. In cities with extensive tax foreclosures and few, low bids for their liens or foreclosed properties, these properties can be transferred to Land Trusts, which as non-profits, can hold the property, perhaps have an interim beneficial use such as urban farming, and eventually put the land into a redevelopment project.

Notes

1 www.sior.com/about/sior-history#undefined2.
2 Cunningham, Storm, "North Carolina Town Tries Crowdfunding the Restoration of a Historic Mill," October, 2015.
3 Kolko, Jed, "Where the Empty Houses Are," *The Atlantic*, November 6, 2013. www.theatlanticcities.com/housing/2013/11/where-empty-houses-are/7497/.
4 Klein, Robert, "Community Blight Solutions," July 2015.
5 www.nahb.org/assets/docs/ises/HowtoCashInMasterSessionAgenda_20140113105053.pdf.
6 Perry, David C., Wim Wiewel and Carrie Menendez, "The University's Role in Urban Development," (Land Lines Article) *Enclave to Anchor Institution*, Lincoln Land Institute, Landlines, July 2009.
7 Live Work Play LOHAS.
8 Newman, Oscar, *Defensible Space*, Macmillan, 1972 and other writings.
9 Kilpatrick, John A. and Bill Mundy, "Appraisal of Contaminated Land in the United States," *Journal of the Japan Real Estate Institute*, October, 2003.
10 Garreau, Joel, *Edge City: Life on the New Frontier*, Anchor Books, 1992.
11 Cocoziello, Peter at New York University, December, 2015.
12 Porter, Michael, "On Competition", *Harvard Business Review*, 1980–1995 and many other works.
13 www.brookings.edu/research/one-year-after-observations-on-the-rise-of-innovation-districts/.
14 www.planningreport.com/2014/07/24/demise-tif-funded-redevelopment-california.

15 Yang, Jennifer, "The Birthplace of BIAs Celebrates 40 Years." *Toronto Star*, April 18, 2010.
16 Whyte, William H., *The Social Life of Small Urban Spaces*, Project for Public Spaces, 1980.
17 www.montgomerycountymd.gov/OLO/Resources/Files/2015_Reports/OLO%20Report%202015-7.pdf.
18 Hersh, Barry, "Real Estate Tax Policies and Brownfield Redevelopment," Lincoln Institute of Land Policy, Cambridge, MA, 2002.

9 Megaprojects

Barry Hersh

This chapter discusses the growing phenomena of large-scale redevelopments that involve substantial acreage for urban area, multiple buildings, transportation and other infrastructure, millions if not billions of dollars. Such projects are often in large, gateway cities, almost always public-private partnerships with government helping by providing land, infrastructure or financing as well as regulatory approval. Megaprojects may or more likely may not include a megastructure, a single large connected multiuse building a concept that was first popularized in the 1970s. Today, it is more likely that, learning from projects such as Battery Park City, the large project uses the existing street grid and are connected to the urban fabric, rather than separated from the existing city.

Megaprojects very often involve all or at least most of the key aspects of urban redevelopment. There is almost always some level of environmental remediation. Often one or more historic buildings or contextual design is required. Public amenities, parks, waterfront promenades and cultural facilities are very often key aspects of these large projects. These projects in North America also generally include affordable housing and community facilities, as well as multiple types of retail, office and residential development. In several cases, competitions were held by the public agencies to select the developer and design team with the best proposal in terms of urban design, public benefits and financial capability

Financing for megaprojects usually involves both private and public funding. Traditional real estate investors, such as pension funds and insurance companies, often participate, but large equity and hedge funds have also participated. Government support can come in multiple funding streams for infrastructure, land, affordable housing or public amenities and are often not direct, on-balance sheet from governments, but rather Industrial Development Bonds or TIF. The low interest rates from 2008 onward have also helped projects. The availability of capital, from traditional real estate investors such as insurance companies and wealthy individuals, to newer sources including hedge funds, sovereign funds, EB-5 visa financing and crowdfunding, have all helped fuel megaprojects.

Megaprojects reflect many changes that are impacting real estate, finance and communication. One driver towards large projects is the cost and length of the governmental review process, only a large project can justify an extensive Environmental Impact Statement, complex financing and time-consuming community engagement. Projects that are based upon public facilities, such as new mass transit stations, often become public-private partnerships with contractual commitments in terms of what the developer and the government agencies each must do and in what sequence. Often government financial support, such as loan guarantees, tax increment financing (bonds based upon future increase in tax revenues) and provision of public facilities are part of such public-private agreements. The private

Case study: Atlantic Station, Atlanta, Georgia

A 138 acre brownfield site located in the midtown heart of Atlanta, Atlantic Station was planned to provide a total of fifteen million square feet of new mixed-use development. The Post Office gave the project its own zip code, 30363, and it certainly qualifies as a megaproject. Jacoby Development contracted to acquire the former Atlantic Steel mill in 1997, and with financing from AIG Global Real Estate, the first buildings were occupied in 2003, overcoming numerous challenges in a relatively few years. It is not coincidental that AIG was and continues to be among the largest providers of environmental insurance in North America, though the environmental insurance business name was changed to Chartis, and efforts to be a real estate investor in brownfield projects largely ended after AIG's troubles in the 2007 fiscal crisis.

Atlanta has been called a poster child for sprawl,[1] and this large-scale redevelopment within walking distance of downtown was intended to provide an in-city option. However, access to the site was limited, and the road improvements needed were held up by air quality guidelines limiting new road construction – even though a project goal was to reduce auto dependency. (A similar issue arose regarding the planned Willet's Point Development in Queens, New York).[2] The road projects, notably the $100 million 17th Street bridge, were approved under the innovative EPA XL program, which allowed greater flexibility for smart growth projects, and US EPA brought in noted new urbanism designers Andres Duany and Elizabeth Zyber-Platek as consultants to review the plan.

The City of Atlanta and the State of Georgia have been consistent supporters of Atlantic Station, at least in part to counter sprawl criticisms. The site was rezoned, after negotiation, to allow the relatively high-density development. A special taxing district was formed, allowing the use of a type of TIF to help fund the extensive infrastructure costs. The Georgia Department of Environmental Protection supervised and approved the over $10 million cleanup of the property. The State also substantially financed the major road improvements, notably the new multilane 17th Street bridge noted for its yellow color. Despite the word station in its name, Atlantic Station does not have a MARTA (Metropolitan Atlanta Regional Transportation Authority) stop but now runs a shuttle to a stop close by. There is also bus service and an active proposal to relocate that station so it is within walking distance to the project.

Atlantic Station has three main areas. The District, bounded by major corridors, is the main retail and office district. Primarily a retail developer, Jacoby chose to design a pedestrian-oriented open-air mall with underground parking, a neighboring hotel and a nearby major office component. The Commons area of Atlantic Station is primarily residential condominiums and features a large man-made retention pond. The third area is called the Village and is also primarily residential and includes an Ikea. Other builders, including Beazer, constructed some of the residential buildings.

Atlantic Station consistently features its environmental qualities using the "Live, Work, Play" marketing theme also employed at other large redevelopments including Stamford's Harbor Point. With a large multiplex cinema, outdoor gathering places and central location, Atlantic Station has become a major entertainment hub, hosting numerous major events throughout the year. Virtually all the buildings have achieved LEED certification at various levels from the US Green Building Conference. The project preceded the

LEED Neighborhood Design standard, but buildings did receive LEED points for being part of a brownfield redevelopment.

There are two artifacts of note in Atlantic Yards. The first is an Atlantic Steel mill smokestack, relocated and readily understandable as a symbol of past history. The second is called the Millennium Arch, an only somewhat scaled down version of the Arc de Triumph. Originally intended for a L'Enfant Circle in Washington DC, it was brought to Atlantic Station by its main promoter, Atlanta classical architecture advocate Rodney Mims Cook, who proudly proclaimed it "the first such (classical) project built since the Jefferson Memorial."[3] There is 12,000 square feet of museum space in the structure that is open to the public.

With an estimated market value in excess of $2 billion, Atlantic Yards is among the largest and most successful urban brownfield redevelopments in North America and an important part of Atlanta's urban fabric.

Case study: Manhattan West Side, the High Line and Hudson Yards

It is hard to imagine a larger scale, more dramatic urban redevelopment than what is still unfolding on the West Side of Manhattan, along the Hudson River. What had once been a set of working piers, contaminated industrial districts, train tracks and gritty neighborhoods such as the Meat Packing District, Chelsea and Hell's Kitchen are being transformed into a multibillion dollar redevelopment, including thousands of residential units, cutting-edge office, ultra-fashionable retail and two new world-famous parks. Hudson Yards alone is a megaproject, with an estimated value of $28 billion, but it is also tied to High Line, whose restoration and new use as a public park has in part led to additional billions of development dollars and transformed several neighborhoods on Manhattan's West Side.

In the 1970s, all of the cargo and most of the cruise shipping had moved from the West Side piers. A proposal to lower the deteriorating West Side Highway and build over it and out into the Hudson was defeated by a combination of environmental and community activists. Fifty years later, a combination of public and private decisions has led to an urban transformation.

It took several tries for Hudson Yards to become reality. Plans for redeveloping the West Side of Midtown Manhattan, including building over railyards, go back over fifty years. There were various proposals for a stadium for the New York Jets football team, tied to a bid for the 2012 Olympics, relocation of the nearby Javits Convention Center and more. The first RFP issued by the MTA (Metropolitan Transportation Authority) in 2007 and featured a "starchitect" competition, which was won by Tishman Speyer – but it soon fell apart in part due to the financial crisis. The second RFP resulted in the project by a partnership of Related Companies (a major New York-based developer) and Oxford Properties (the Ontario, Canada, pension fund) and a master plan by KPF architects. The original 28 acre site, much of it a platform built over the railyards, has been expanded by very expensive land acquisition, while other major developments such Manhattan West by Brookfield (the Canada-based successor to Olympia & York) as well as projects by the Moinan Group and Tishman Speyer. Hudson Yards is still growing, with much of the largely office eastern portion complete or under construction, while the largely residential western portion still awaits construction of its platform over the railyards. The recently completed northern-most stage of High Line wraps around three sides of the massive Hudson Yards development and was built in partnership with the Oxford/Related team responsible for Hudson Yards. An enormous platform now covers half the existing

Megaprojects 139

Figure 9.1 High Line as it approaches Hudson Yards
Source: Photograph by T. Lawrence Wheatman.

Figure 9.2 Hudson Yards with High Line, under construction
Source: Photograph by T. Lawrence Wheatman.

railyards used by Amtrak, the Long Island Rail Road and the New Jersey Transit. The first office building could start earlier as it is built on terra firma abutting the train yard and is now partially occupied and owned by Coach; other users include SAP and L'Oréal. The extension of the Number 7 subway line to serve the area has opened. Hudson Yards has been described by *Forbes* as "America's Biggest Real Estate Project ... Ever";[4] to include over 5,000 residences, 5 office towers each over one million square feet, over 100 retail shops including the first Neiman Marcus in Manhattan, plus over 14 acres of open space, a culture shed and many other amenities. Oxford/Related was selected after a complicated competition, but the plan was modified to fully incorporate the High Line.

The signature West Side project is the High Line, the redevelopment of an elevated freight railroad into a 1.45 mile linear city park in almost the same time frame as Hudson Yards. In 1999, Friends of the High Line, led by two young residents, reacted to the Giuliani administration proposal to demolish the rail viaduct (a small southern section had been taken down earlier), held a design competition and put forth a redevelopment plan. With the active backing of several prominent New Yorkers (some of whom, such as Barry Diller and Diane Von Furstenberg, owned nearby property) and a stellar design, Mayor Bloomberg was convinced to reuse the High Line. High Line Park was built in three stages, the first opening in 2009, the second in 2011 and the last in 2014. The park cost in excess of $400 million with private sources, including early donations and later the Hudson Yards redevelopment participation (encompassed on three sides by the northern portion of the High Line) contributing a significant portion of the cost. The High Line has become an extraordinary attraction, with over 4.8 million visitors, both locals and tourists, in 2014.

Perhaps even more dramatic has been the explosive urban redevelopment centered on the High Line, totaling over $35 billion, huge even by New York standards. There have been dozens of residential projects, most luxury market rate rentals and condominiums, designed by world-class architects including Jean Nouvel and Zaha Hadid. There has been office development, starting with the Frank Gehry-designed IAC headquarters, plus the Google $1.9 billion acquisition of former Port Authority of New York and New Jersey industrial buildings. The galleries in Chelsea have multiplied, as have high-end boutiques, plus a new Whitney Museum has been built. Some were concerned about the space under the High Line; the existing Chelsea Market was reinvigorated and is to be expanded, including additional restaurants and boutiques. The timing was fortuitous, as the first phase of the High Line was opening, New York real estate was starting to recover from the 2007 financial recession. The West Side began to capture a significant portion of the surging high tech and media markets.

As might be expected in a heavily built-up community, there were important land use and zoning aspects of the redevelopment. In 2005 the New York City Planning Commission created the West Chelsea Special Zoning District using transfer of development rights to compensate owners whose development opportunities would remain limited due to the rail line remaining and to protect the light and viewshed of the High Line. First the new zoning district allowed owners of property under the High Line (which was legally an easement) to transfer development rights in a wider than normal area – vastly increasing their value. Additionally, the new zoning was an effort to control new development, keep it far enough away so that the High Line views, light and air were protected – and it steered high-rise development mostly to the east, towards the center of Manhattan Island while maintaining views of the Hudson.

There has been criticism that the High Line redevelopment, despite being a park open at no charge to all, has resulted in gentrification, especially given the many $5 million and up condominiums. Some of the apartment projects near the High Line and in Hudson Yards do include affordable units. Nearby public housing, low-cost cooperatives and rent-stabilized apartments remain. It is clear that the once gritty industrial area with moderately priced older housing, after an influx of artist and gallery pioneers, is now much more upscale.

The Midtown West Side of Manhattan was not particularly close to subways by New York standards; it was a fifteen to twenty minute walk from some areas, though there are also buses. Most importantly, Mayor Bloomberg who was the champion of west side redevelopment, committed funding, and the Number 7 subway line has been extended to the northern Hudson Yards area, which is also close to the High Line and Penn Station regional transit hub. The West Side development now certainly qualifies as transit oriented.

Most of the sites on the West Side in the overall roughly 500 acre development area were originally landfill and contaminated by former industrial facilities, gas stations and Manhattan's last scrapyard. The cleanup of the High Line itself received a $200,000 EPA Brownfields grant via the City of New York. While there was no overall brownfield planning program, over forty sites in total were remediated under the supervision of either state or city environmental agencies, which certainly constitutes an area-wide brownfield cleanup. Seven projects have used, and others sought to be eligible for, New York State's lucrative Brownfield Tax Credit programs. Twenty-six other sites went through New York City's voluntary cleanup program, which offered far less financial incentive but far quicker processing. The vast majority of the contaminated sites in this area were remediated; generally polluted soils were removed or encapsulated. It is hard to measure, but this is believed to have contributed to the overall environmental quality of the nearby Hudson River as well as the West Side of New York City.

A recent hyperspectral imaging study conducted by the New York University Center for Urban Science and Progress (CUSP) showed that the most common air contaminant was ammonia, coming largely from large air conditioning chillers.[5] There were only small traces of methane or other pollutants associated with contaminated sites. The water quality of the Hudson River has also improved dramatically due to many efforts by government at the federal, state and local level, as well as by non-profits such as Riverkeeper and League of Conservation Voters. Arguably, a small portion of the improvement resulted from the riverside brownfield remediation. Hudson Yards, working with the NYU CUSP (another initiative supported by Mayor Bloomberg) is utilizing an extensive "big data" system that monitors not only energy, water and sewer flows, but also pedestrian and vehicular traffic to improve efficiency, sustainability and resilience. In addition to the High Line, the other improved open space is Hudson River Park, built in sections connecting with Riverside Park to the north on the Upper West Side and Battery Park City promenade and Battery Park itself to the south, now part of a long-term vision to provide waterfront access to the Hudson and all around Manhattan Island. There have been numerous park amenities, some a part of Hudson Yards, as well as a recent additional new pier park proposed by some of the High Line funders.

The West Side of Manhattan is a case study of urban redevelopment writ not just large but humongous. Clearly this development requires the huge and thriving New York real estate markets. Yet the basics remain the same; a valuable urban location, good transportation including extended mass transit, a public-private partnership and community support. It is an almost waterfront site (State Route 9, formerly called the West Side Highway separate the site from the Hudson River), with a view of the Hudson, and a significant portion or urban redevelopments have water amenities. These multibillion dollar projects used a

Figure 9.3 Hudson Yards and High Line Spur
Source: Photograph by T. Lawrence Wheatman.

whole range of financial sources, mostly private and notably include EB-5 visa financing. There were serious environmental issues to be overcome, as many if not most urban projects experience. The new buildings, following New York City's strict energy codes are energy efficient, while the entire area including parks proved resilient even in the face of 2012 Superstorm Sandy. While the scale is enormous, how this area was transformed is similar to many smaller projects, with the government helping to provide part or all the site and infrastructure, the developer pushing through a complex, multifaceted process, attracting users and financing, a well-orchestrated design effort and gathering community support and providing amenities.

Rebuilding Detroit

If ever there was a city that needed urban redevelopment it is Detroit, 2012, as described by Rod Stevens in the opening chapter. The city declared bankruptcy, population was fleeing, and despite some noteworthy efforts especially downtown, the city was put into receivership. Edward Glaeser notes in *Triumph of the City*[6] that the massive megastructure, somewhat urban renewal style, of what was first called Renaissance Center, later RenCen, was not the savior GM and other Detroit leaders hoped it would be, a fortress that turned its back on the waterfront, and did little for downtown and less for any neighborhood. Most of all, it failed to address the basic economic changes in the auto industry and the region. Some of the numerous downtown revitalization efforts, such as the award-winning Horace E. Dodge and Son Fountain and the relatively new Tiger Stadium remain strong assets for the future. Detroit, which used to be a Midwest epicenter for a thriving industrial economy, has long been suffering the effects of the decline of the automotive industry. The city had been the 4th largest urban region in the United States during the mid-twentieth century, but it has now been reduced to only the 18th largest U.S. city.[7] Detroit is currently aggressively pursuing redevelopment of neglected or blighted land, with some success downtown and also in neighborhoods in part via a Land Trust. It is also trying to address high unemployment through job training and enticing a greater diversity of employers, including technical and creative businesses. Rebuilding Detroit is still very much a work in progress. There are so many ongoing and new initiatives, each reflecting one or more key aspects of urban redevelopment.

Detroit remains the most populous city in Michigan and is located on the Detroit River in the northern region of the state, adjacent to the U.S.-Canada border. The region is still second only to Chicago as one of the largest economic regions in the Midwest, and approximately half of Michigan's population resides there.[8,9] When the auto industry expanded in the 1950s and 1960s, Detroit was home to just under six million people, but after the auto industry declined, many jobs were lost and residents moved to suburbs. The City experienced more than a 60 percent decrease of the population, and today there are fewer than 750,000 people. Population decreased by a quarter during the decade of 2000 to 2010.

Overcoming high unemployment and prevalent land vacancy have been severe challenges for Detroit. Efforts to once again position the City as a Midwest economic center include very large-scale, and sometimes unconventional, urban revitalization plans. Redeveloping unused or historic buildings as entertainment venues has been a recent success for Detroit; for example, the Theater District has seen resurgence as a popular commercial area, and riverfront attractions are also popular. But the City still has a significant way to go in revitalizing most of its neighborhoods and abandoned properties. Detroit declared bankruptcy in 2013, with the filing being the largest municipal bankruptcy case in U.S. history.[10] The bankruptcy plan was approved, and the City of Detroit exited Chapter 9 Municipal Bankruptcy at the end of 2014.

According to a parcel survey from 2009, 25 percent of residential lots in Detroit were either undeveloped or vacant, and 10 percent of developed housing was unoccupied.[11] One resolution has been to designate unused land for other purposes – notably urban agriculture – while also relocating residents from more isolated neighborhoods to those that are denser. Detroit has demolished several thousand abandoned houses and properties to date to further alleviate vacancy issues. What to do with the land once the abandoned houses have gone is challenging everyone's creativity. Hantz Woodlands, which bills itself as the world's largest planned urban farm, wants to plant at least 15,000 trees in eastern Detroit. Others say the city should create big fruit and vegetable farms and become a gastronomic hub such as Traverse City in Michigan, which successfully markets itself as farm-to-table foodie heaven. Detroit is frequently described as a "food desert" where people cannot shop for fresh, wholesome food. Urban farms could change that within a couple of years.[12] Hantz Woodlands commenced in 2013 and involves the redevelopment of more than 1,500 vacant, previously city-owned lots on the east side of Detroit into an "urban forest." The 140-acre redevelopment includes the razing of fifty derelict properties, cleanup of garbage, planting of 15,000 trees and maintenance of the landscape. The redevelopment had been controversial and was approved by City Council in a narrow 5–4 vote, as critics of the plan

144 B. Hersh

claimed it would increase land values by depleting inventory from the available real estate market and allocating to a specific land use of agriculture only. The owner of Hantz Farms, the buyer and developer, agreed but argued that such a move would be beneficial to Detroit and would help to "stabilize the market." He also further claimed that the goal of this enterprise would be to eventually create jobs in tree maintenance and the cultivation of new plants. Hantz Woodlands is open to the public and is scheduled to complete in 2017.[13]

The City is also pushing efforts to establish and execute redevelopment plans for other distressed neighborhoods; there are currently six working neighborhood revitalization plans for areas throughout the City, and some private financing has been committed to assist the City with these plans. A seventh plan, the Far Eastside Plan, covers 1,200 acres which have been razed for the purpose of new large-scale development. In 2008, the City had committed $300 million raised from bonds, for not only neighborhood revitalization, but also to generate jobs.[14,15]

The Detroit Riverfront Conservancy is also developing and managing revitalization on Detroit's "International Riverfront." This area ranges from the Ambassador Bridge to Belle Isle Park in the downtown area of the City and includes several noteworthy parks, the famous Dodge Fountain, restaurants and shopping, high-rise commercial and residential buildings along the Detroit River. Detroit has always had its advocates, such as Harriet Saperstein who worked tirelessly as a planner with a special mission to improve the riverfront.[16] The Conservancy was established in 2003 and opened a new attraction in 2007.

Figure 9.4 Hudson Yards construction 2016

Source: Photograph by T. Lawrence Wheatman.

Recently, seven design concepts have been proposed for East Waterfront, near downtown and nearly opposite the famous Belle Isle Park.

The new Detroit Red Wings arena is under construction, part of a plan for the redevelopment of forty-five city blocks in downtown and the adjoining midtown areas, and a new mixed-use neighborhood in Cass Park. A public-private consortium started in 2007 is continuing with plans for 3.3 mile M-1 streetcar rail that will run along Woodward Avenue and will link downtown and midtown with twelve stops including one at the new Red Wings arena. The M-1, which received considerable federal support after the bankruptcy, is under construction and now expected to open in 2017.

Infrastructure improvements are an important part of rebuilding Detroit. Replacing streetlights with brighter (double the wattage), energy efficient bulbs not only enhances public safety but encourages more pedestrian and retail activity and is actually less costly and improves public safety. A bond issue for $185 million by Detroit Lighting Authority is paying for more than 56,000 new bulbs.[17] More than 35,000 broken streetlights have been replaced.

Affordable housing is a key component of any large-scale redevelopment. Detroit has many older buildings that can be creatively readapted for housing as well as other uses. Due to abandonment, there is still too much vacant land in the city, but it does provide opportunities for new affordable housing development. While from 20 percent to 40 percent affordable housing is encouraged in new development, there is not a formal inclusionary zoning requirement.

More recently, there has been a surprisingly strong response to the bankruptcy, though it is probably too early to declare victory. Demands by creditors that the city's assets, including the museum's art collection be sold have largely been rebuffed and resolved. Local companies, national organizations and both the state and federal governments have joined a new mayor in efforts to rebuild. Under Diego Rivera's murals of "Industry" at a dinner for supporters of the Detroit Institute of Arts, Anne Parsons, who runs the Detroit Symphony Orchestra, remarked how much the mood in the city has changed. "A few years ago, everyone would have been pessimistic at a similar gathering," says Parsons. "Today the happily chatting visitors seem to feel the worst is behind them."[18]

Mayor Mike Duggan considers himself a turnaround specialist. In his first state-of-the-city message, he talked about the first balanced budget in over a decade as Detroit recovers from the largest municipal bankruptcy in U.S. history. City services are said to be somewhat improving: examples cited include that the number of ambulances has doubled and their response time has dropped from eighteen minutes to an average of eleven (the national target time is eight). More than 220 parks have reopened with the help of churches and community groups. Detroit no longer has the highest murder rate in the country.[19] Although it is still a dangerous city, the rates of murders, robberies and carjacking's are falling. Duggan also cut residential property tax assessments by 5 percent to 20 percent, based upon current market conditions and encouraging residents to maintain and stay in their homes. The city is still demolishing 200 derelict houses every week.[20]

Many, including the mayor, talk about much more needs to be done, and the recovery seems mainly focused on the greater downtown, where many earlier publicly supported amenities and developments, including several sponsored by local billionaire Dan Gilbert, are occurring but focused on a small area representing roughly 5 percent of a large city. There is a Motor City Match of $500,000 quarterly for the coming five years to encourage innovative entrepreneurs, and much of the initial focus has been on business start-ups in shared space in the downtown and midtown core. Major infrastructure systems, including public transport and the public schools have not noticeably improved. Detroit remains one of the rust belt's shrinking cities, but it does have the advantage of having several universities in the area or nearby (or with satellite facilities), including Wayne State, University of Michigan, Lawrence Technological University and Michigan State, as it, like many other places, tries to replicate some of the technological growth aspects of Silicon Valley.

> Ted Serbinski is the sort of person Detroit would like more of. He moved there three years ago from San Francisco, thinking the city might become a Midwestern Silicon Valley and wondering how it could use its special position so close to the three big carmakers. He is now the managing director of Techstars Mobility, which launched in December in partnership with Ford, Verizon, a telecoms company, and Magna International, a car-parts supplier. The new company is an incubator: it will invest in ten startup companies each year that link up technology and cars. The first ten companies will be announced in June. The hope is that one of them may prove as successful (though maybe not as controversial) as Uber, which links taxi passengers to drivers through a smartphone app.[21]

The founder of Quicken Loans and a leader in rebuilding Detroit, native Dan Gilbert, has also created Rock Ventures LLC Bedrock Real Estate Services LLC which owns more than seventy-five properties and more than 12.5 million square feet of space including parking, mostly in and around downtown. According to his own figures, Gilbert has invested more than $1.8 billion in the city,[22] and many of his Quicken Loan operations are now located in downtown Detroit. One interesting venture is the Madison building, a center for the growing technology sector. There are also very new tech companies (some of them short-lived) that rent space and work side by side in a bright, open-plan area with a red popcorn maker and a gleaming coffee machine. This is also the home of Detroit Venture Partners (DVP), Gilbert's technology-investment fund, which is now attracting more conventional investors. Another Gilbert project is a $10.3 million plan to turn a vacant Capitol Park building into 5,500 square feet of first-floor retail space and twenty-five residential units on the top five floors has received approval for $1 million in state financing. The project is to redevelop the building Gilbert owns at 1215 Griswold Street, north of State Street. Gilbert will provide all of the upfront financing for the project with the expectation of reimbursement from $1.65 million in federal historic tax credits; the $1 million state loan and $2.85 million will be financed with traditional debt, according to the memo. The remaining $4.8 million would be funded through owner equity. Robin Schwartz, public relations director for Bedrock, said construction is expected to be complete in the early fall.[23]

Michigan's Community Revitalization Program provides grants, loans or other economic assistance of up to $10 million to projects that will revitalize regional urban areas, act as catalysts for additional investment in a community, reuse brownfield and/or historic properties and promote mixed-use and sustainable development. Lansing-based The Christman Company is the general contractor on the project. Detroit-based architecture firm Kraemer Design Group PLC is the project architect. The loan is the first Michigan incentive received by Bedrock. The redevelopment of the former Bethlehem Steel plant in Bethlehem, Pennsylvania, described in Chapter 2 has been cited as a model of imaginative, economically varied and sound planning of a former heavy industrial facility, such as a range of new and adaptive reuses that include new industrial facilities, public event space and a hotel and casino.[24]

Brownfield tax credits have been used in numerous projects including some of Gilbert's, such as the redevelopment of several buildings on Woodward Avenue and the former Federal Reserve Building downtown, according to the memo, that was purchased in October 2013 for a local developer.[25] Among the many concerns is whether the energy and enthusiasm that is palpable downtown on the riverfront and midtown can expand to the rest of the city. Property developers, which include some established locals, a few start-ups and international developers, are still mainly focused on the center of the city, where prices have increased the past few years, unlike many neighborhoods. In 2012, Jordan Wolfe and a partner bought Claridge House, an apartment building downtown, for $750,000 and renovated it. Today, he says, it would cost between $2 million and $3 million. All its forty-five apartments are rented out, and there is demand for more. For the first time ever, Detroit has waiting lists for residential space downtown. There have also been proposals to reform Detroit's property tax system, even beyond the mayor's recent reduction for homeowners. A study supported by the Lincoln Land Institute made numerous proposals, including refining assessments, targeting tax abatements and using a land value tax base.[26]

Property investors are slowly branching out into districts such as Eastern Market, an area bordering on midtown with a huge outdoor flower and food market, and Corktown, just west of the center. Corktown was among the Detroit neighborhoods that recently received grants from the Kresge Foundation, based in a nearby suburb. Goldman Sachs is supporting entrepreneurs with a $500 million program.[27] New eateries are popping up, bringing a variety of cuisines downtown and to the stronger neighborhoods.[28] Yet large parts of the city remain a wasteland. That is why Duggan is so keen to remove the more than 80,000 derelict buildings, which the Blight Removal Task Force, a public-private partnership, wants to demolish. "Though stable neighborhoods still exist, they are overshadowed by the city's blighted areas. This negative perception ... has been a factor in hindering the city's growth," said a representative from the Federal Reserve Bank of Chicago in a recent report. The federal budget package approved at the end of 2015 will, at the urging of Michigan Congressman Kildee, allow more of the Hardest Hit Fund, intended and still used for mortgage relief to now be used more for demolition and blight removal, especially in the Midwest and Detroit.[29]

Among the more interesting revitalization efforts in Detroit are those by homeowners themselves. "Blot" is a term created to describe a homeowner who purchases adjoining, now vacant, properties.[30] Detroit and many other cities offer "first dibs," some advantage in timing and/or price for acquisition of an abandoned property to the adjoining property owners. Homeowners are protecting their property and controlling their surroundings, and may use their expanded property as allowed by zoning for gardens, lawn, better driveways and parking. Some owners may also hope for appreciation and potential sale or to participate in future redevelopment.

There is much to be positive about in Detroit's future growth, especially now that the City has successfully exited bankruptcy. In addition to the concentration on commercial growth, enhancement and additions to existing infrastructure and burgeoning entertainment centers, the many small independent businesses and strong arts community contribute to the unique character of Detroit. Overall economic redevelopment, local culture and the attraction of affordable space within a City show promise of continued growth and promise for Detroit.

Notes

1 Burns, Rebecca, and Report: Atlanta is the most sprawling big metro in the U.S. quoting a study by Smart Growth America, April, 2014.
2 De Sousa, Christopher and Lily-Ann D'Souza, "Atlantic Station, Atlanta, Georgia: A Sustainable Brownfield Revitalization Best Practice," 2013. Institute for Environmental Science and Policy, University of Illinois.
3 Dewan, Sheila, "An Elaborate Arch, an Opaque Significance," *New York Times*, April 29, 2009.
4 Sorvino, Chloe, "Hudson Yards, America's Largest Private Real Estate Development, Opens First Building," *Forbes*, May 31, 2016.
5 Ghandehari, Masoud, "Cities in the Eye of the Snake, Hyperspectral Imaging of the Westside," New York University Center for Urban Science and Progress, July 2015.
6 Glaeser, Edward, Op cit.
7 www.citymayors.com/gratis/uscities_100.html.
8 "Economic Growth Widespread Across Metropolitan Areas in 2014," *Bureau of Economic Analysis*. United States Department of Commerce.
9 "2010 Census Interactive Population Search," U.S. Census Bureau.
10 *The Detroit News*, October 13, 2013.
11 Detroit Parcel Survey.
12 "Green Shoots," *Economist*, May 1, 2015.
13 www.citylab.com/design/2013/10/140-acre-forest-about-materialize-middle-detroit/7371/.
14 "Next Detroit at the Wayback Machine," *City of Detroit*, May 2, 2008.
15 "Community Development at the Wayback Machine," *DEGA*, February 4, 2008.
16 Waterfront Center Conferences 1990 to 2014, www.secondwavemedia.com/metromode/features/saperstein88.aspx.
17 www.freep.com/story/news/local/michigan/detroit/2015/11/12/detroit-street-lighting-project-update/31850609/.
18 "Green Shoots," *The Economist*, May 1, 2015.
19 *The Detroit News*, October 2, 2015.
20 http://archive.freep.com/article/20140127/NEWS01/301270081/Duggan-property-tax-assessments.
21 "Green Shoots," *The Economist*, May 1, 2015.
22 www.freep.com/story/news/local/michigan/detroit/2014/09/28/downtown-m-rail-arena-bridge/16343825/.
23 "Green Shoots," *The Economist*, May 1, 2015.
24 www.washingtonpost.com/business/case-in-point-in-bethlehem-pa-a-road-map-for-detroit/2013/07/25/a744d070-f3dc-11e2-a2f1-a7acf9bd5d3a_story.html.
25 CoStar, 2013.
26 Sands, Gary and Mark Skidmore, "Detroit and the Property Tax Strategies to Improve Equity and Enhance Revenue," Lincoln Land Institute, 2015.
27 www.freep.com/story/money/business/michigan/2015/08/06/small-business-grad/31210193/.
28 Guarino, Mark, "One of the Country's Poorest Cities Is Suddenly Becoming a Food Mecca," *Washington Post*, January 5, 2016.
29 "Congress Passes Detroit-Led Demolition Measure, Great Lakes Protection and More," *Engineering News Record Midwest*, December 18, 2015.
30 Interboro Partners, Tobias Armborst, Daniel D'Oca, Georgeen Theodore, Christine Williams, "Improve Your Lot," 2008 and 2011.

10 The urban redevelopment process
Putting it all together

Barry Hersh

Anyone who says there is only one thing that matters in a redevelopment project, be it "location, location, location" or urban design or financial structure, does not fully understand the process. It is always a combination of factors including location, finance and design, community engagement, transportation, remediation, resilience, market, infrastructure, preservation and amenities. There is no single model for urban redevelopment. There are far too many unique examples of success and failure for one book, but there are connecting threads. One of the overarching challenges of having so many complicated aspects is that it is easy to lose sight of how each of these individually affect not just cities and communities but each family and individual who will live or work in the community.

The urban redevelopment process can be seen in some ways as a variation on the real estate development process, described in detail in the widely used Urban Land Institute textbooks by Miles et al.[1] and Peiser,[2] among other works. However, each step in the redevelopment process differs from traditional development. While a private developer may typically initiate an urban redevelopment project for profit, the community and government agencies are also likely to play key roles from the very beginning, including the initial concept, to achieve shared goals.

The due diligence on an urban redevelopment site will likely include more about local politics, environmental contamination and the condition of existing infrastructure. Gaining control of an urban site is less likely to be a straightforward purchase and sale, and more likely to include an RFP (Request for Proposal) from a municipality or local development organization, assemblage of multiple parcels and often will feature extensive negotiation of community benefits. Early market analyses may be performed to help frame the RFP and inform the community, but developers are still likely to do their own and more targeted market studies later. Community engagement and expectations are often crucial for both approvals and marketing. There will likely be distinctive building architecture, often featuring historic preservation or new urbanist features as well as urban design. Context, community style and amenities will be crucial as the initial design concept is refined. Designs that provide green infrastructure, ecological soft edges to improve resilience, may also be an opportunity for walking paths that promote healthy activity. Creative use of zoning, site plan approvals, public-private partnerships and the land use process can help support and assure community support for a redevelopment. Affordable housing programs may be utilized to avoid displacement of current residents. Financing will likely include equity and mortgage debt, perhaps mezzanine debt or preferred equity, and also public "gap" finance such as from TIF or revenue bonds or tax credits for historic preservation or low-income housing. There may also be other government gap financing, for remediation, energy efficiency, resilience or community facilities.

Urban redevelopment takes a champion (not just a developer) – every redevelopment project has a person or persons who have the vision, put the pieces together and persevere. The champion(s) can be a developer, a neighborhood activist, a designer or a public official. Without a champion, even a good and much needed redevelopment concept will fall by the wayside.

The people who put together urban redevelopment projects are special; they care about community, history, environment and more than just return on investment. It takes grit, the ability to overcome obstacles and persevere. Sometimes what first seems an outlandish idea, turns out to be creative. The first proponents of reuse of an abandoned train bridge in Poughkeepsie for pedestrians were called crackpots, but after support grew and options faded, a change in perception led to state support and use of the Rails to Trails program that, a decade later, brought about the now very popular Walkway over the Hudson. While the macro trends from Fed policy to the Rise of the Creative Class set the stage, each urban redevelopment is different and finds own path. One concern about current federal policy is that even champions need a place to stand – so that funding for the redevelopment professional is important. Finding project funding, from private and public sources is demanding, but a good project will find support. Long-term funding for positions, for those who carry projects forward, is often harder to secure.

All urban redevelopment has to deal with the context of the urban fabric, making a city not just bigger and newer, but better for its inhabitants. Urban redevelopment requires thoughtful design that is unique to that place, though not necessarily iconic architecture. While almost all urban sites have some remnants of former uses, only some projects require extensive environmental remediation, but promoting a modern, healthy lifestyle is always beneficial. Transportation and infrastructure are systemically crucial, but not every project requires major new facilities. Urban redevelopment looks at existing historic elements to connect and build upon, but only some involve formal historic preservation of landmark buildings. Job creation beyond construction, including new approaches such as innovation zones and incubators, are often part of the project. What the community wants and expects is far more important in rebuilding cities today than in the past. Timing, doing the right project at the right time, often starting with a small, doable portion before taking on large projects can be critical. All the factors matter, but which ones are writ large while others are smaller contributors are different for each location. There is no universal, nor even typical, critical path flow chart – but there are principles. Redevelopment is a reiterative process, plans are constantly refined and small early successes can lead to larger projects.

One illustrative model described in Chapter 6 on brownfields is the Triad remediation methodology, which improves the remediation process by filling information gaps in real time, in the field, so that information is readily exchanged and the remediation can be immediately and constantly refined until completion. Similar computer-based information exchange using modern Building Information Systems is another example of an information feedback system that constantly refines the project. This constant refinement can be considered as a conceptual approach to the urban redevelopment process.

A related concept is illustrated by the HUD–EPA–DOT Partnership, representing an approach that breaks down silos and allows information and decision making to flow across bureaucratic boundaries, and that interactive approach can also be utilized to accelerate a specific project. All these models suggest approaches for developers: the ability to quickly modify based upon actual conditions and coordinating the process among different aspects of these projects.

Most developers multitask, spending their days hurrying from meeting to meeting, dealing with each aspect of the project: design, financing, remediation, public approvals, tenant negotiations, marketing and more. Typically, the developer has a small team and acts as the coordinator, juggling the various concerns and relaying information and too often muddling through, rushing to address the crisis of the day – and possibly creating a new crisis tomorrow. Roger Lewin's book *Complexity: Life on the Edge of Chaos*[3] begins to capture how the process actually works in different arenas; one has to look at the interactions as well as the components. Psychologist Daniel Kahneman's Noble Prize-winning work, as described in his *Thinking, Fast and Slow* is also relevant.[4] Michael Lewis found Kahneman's and his late associate Amos Twersky's approach so compelling and foundational to his *Moneyball*, that he wrote another book, *The Undoing Project*, about the two researchers.[5] Their concepts of how decisions are made, those that are made quickly from the gut and the process for careful analyses of complex issues are highly relevant to real estate and especially urban redevelopment. When scientists consider cities they look for models and algorithms, cities have been said to be like organisms from a cell to an elephant and various types of machines. As cited in Chapter 3, architect Christopher Alexander influenced computer programing as well as design in A Pattern Language and a noted essay, *A City is not a Tree*. Today some use big data, tracking everything from electricity to water and pedestrians to gain insights on how urban systems function. Technologies, ranging from remediation techniques to data analyses can work – under certain circumstances, no one algorithm fits all.

One frequently noted perception is that projects seem to move through the process at what feels like glacial speed, but then as physical work begins, the projects seem to erupt. No matter how many hearings, articles and blogs, there are always those who are suddenly aware and concerned. The well-orchestrated project has built a strong record and good level of consensus to have the support to withstand the generally predictable demands and concerns that will appear as the project approaches reality, as well as external events such as financial and election cycles. The ability to adjust to changed conditions and to coordinate among review agencies can be invaluable. This requires a tight, well-knit team with a wide range of expertise.

Often redevelopment project champions have, in their head and sometimes using software packages, what is effectively their unique critical path method chart that identifies key decisions and benchmarks significant in moving the project ahead. In a complex set of approvals and decisions, can mad fire drills to address immediate crises be minimized, even if never completely avoided? To what degree can the process be codified and made transparent so that various professionals working on different aspects of the project are fully informed and can interact and advise one another to reach decisions? All of the development projects operate behind the curtain, attempting to present a smooth, organized image. It is only when things go wrong that the curtain is pulled aside and failure to address an issue becomes apparent.

The following simplified flow chart illustrates how the early stages of the process are highly complex and relatively risky. Typically, the developer has most control early in the process, when withdrawal is more of an option. As the project proceeds and costs incurred grow, the developer has less effective control, decisions have been made, plans approved and they are harder to modify while withdrawal would become far more painful. Planning and creating a structure that can deal with all of the complexities is crucial; trying to deal with each issue only when it appears is a recipe for failure.

Redevelopments with special features, such as waterfronts, have even additional complexities that are generally best addressed early and then throughout the development

152 B. Hersh

process. From the brownfields perspective, once the developer knows how much the remediation will cost, when it will be done and how liability concerns will be addressed the project becomes more conventional. As the quantitative studies suggest, one way to look at these projects is that resolving environmental concerns restores the property to its full real estate value.[6] Sometimes the environmental resolution must be resolved very early, before the deposit goes hard, which also allows for environmental constraints to inform the design effort. During the design and construction phases, there are chances to gain synergy by coordinating the redevelopment and the remediation. Any institutional controls for remaining contamination have to become part of the long-term operation and costs for the project.

Similarly, waterfront access and amenities have to be thought of from the start and be part of designed and planned operations and often become a key part of the approval agreement with the municipality. Community concerns, whether about typical land use and traffic issues or environmental or waterfront issues, must be addressed directly and are relevant not just when seeking approvals, but throughout the project. While it is impossible to do everything first, it is possible to have a complete early checklist, begin the right solutions, and move the project forward on multiple tracks.

The flow chart also reflects, in a simplified manner, that there is also a financial track as well as design and planning, environmental, marketing and construction tracks, which are all interconnected. Raising the equity to gain site control, incorporating government funds, convincing construction, permanent and possibly mezzanine lenders plus insurers are some of the biggest challenges. For commercial projects, the developer would most likely need some strong tenants, so those negotiations and considerations also get fed into the design and construction aspects of the project.

This type of flow chart does not fully reflect the reiterative nature of development, especially complex urban redevelopments. There is constant feedback, problems, changes and

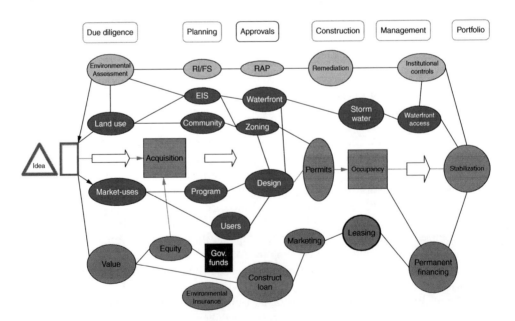

Figure 10.1 The complexity of redevelopment flow chart
Source: Barry Hersh, Waterfront Redevelopment, NAIOP Foundation, 2012.

adjustments. Just as a pilot goes through a very specific checklist before the plane takes off, the project champions need to have their own checklist of what needs to be in place to deal with the many anticipated issues and to help prepare for the unexpected and critical decisions.

Financial feasibility is always paramount on the checklist, but other factors also reappear. As noted for a brownfield project, the contamination has to be identified and the remedial plan understood. Special waterfront and sustainability concerns will permeate the often long and cumbersome approval issues and will often require sophisticated design resolution. The decision system may not be elaborate, but it must be capable of dealing with the number of variables and complexity of the project. Think of a three-dimensional Building Information Management (BIM) system applied not just to building design and construction but to the entire project process.

Another type of approach that accepts, to some degree, somewhat organized chaos is assuring constant communication, rather than to over codify. Can interaction be supported to the extent that the various professionals are in constant communication or can one leader be the hub for decisions? To some degree, the effort to break down silos can be applied to the various internal aspects of the project, as well as among reviewing agencies. Among the approaches that can effectively move things along, the use of charrettes or design workshops is among the most effective. Numerous urban redevelopment projects have utilized such design charrettes to help build community support, and many design firms are new expert at organizing and running such design workshops. Numerous large, multi-aspect projects, including the new World Trade Center, have design, engineering and construction teams sharing space to mimic the charrette approach and facilitate communication.

It is important to use opportunities that gain synergy among these various threads. For example, the big yellow machines working to remediate a site may also be able to start work on foundations. Land trusts that use vacant housing lots provide fresh food, and are an interim use that improve the community and may lead to redevelopment. It is the ability to maximize such synergies that most benefit a project and create a positive disruption to change and improve an urban community.

The historic and current trends clearly tell a story, a confluence of economic, political and market factors have turned to favor urban redevelopment. Urban neighborhoods with their funky shops and walkable character – and modern amenities – are in demand by millennials and investors. Technological changes in everything from transportation and remediation to social communication have enabled previously impossible projects. The overall movement towards social and environmental sustainability strongly favors life and redevelopment in cities and older suburbs. The growth of megaprojects, large-scale redevelopment, is testimony to the financial attractiveness of urban redevelopment. For the first time in generations, market trends are favoring urban redevelopment, but the particular pieces must be brought into place.

Leadership is the indispensable ingredient, starting with the project champion(s) earlier noted. Cities and older suburbs have a history of not only physical but also moral decay. Corruption and crime were important ingredients in the mid-twentieth-century flight to the suburbs. It is important that the seedier past aspects of an urban area be overcome, and those who promote redevelopment can only have long-term success if there is transparency and a sense that the project is being done the right way. It is not that urban redevelopers cannot make a good return on their investment nor that professionals should not be paid for their work – it is that all monies, especially public funds, must be used responsibly and community interests treated with the priority they deserve.

Key factors in success or failure

Preparation and building a team: numerous projects have failed at least in part because the project champion(s) had not done sufficient strategic planning, was spread too thin, lacked capabilities or was too far removed from the project. These projects require multiple skills and lots of hands-on attention. Each of the successful case studies had a strong development leadership team. While each development team is different, it is important to have the skill level necessary in all of the specialties involved. The project champion(s) also needs to have confidence and be in close communication with team members to assure coordination.

Vision and synergy: there needs to be not just a plan, but an overriding concept that guides the project. While there may be multiple goals, affordable housing, remediation, job creation and community amenities, these must be bound together, planned and sequenced with the community, as shown in the Baltimore and Toledo case studies. Mentioned several times, including discussions of Trenton and Stamford, was the opportunity for cost and time savings by actions that accomplish more than one goal. The most common examples are to coordinate remediation and early redevelopment, such as utilities and foundations. Having a strong vision allows for other examples of synergy, coordinated timing of building openings, marketing and even finance, such as going from construction to permanent financing or managing grant cycles.

Project scale: Hudson Yards is a megaproject, as are some earlier projects such as Atlantic Station in Atlanta, Georgia, and Mare Island in Vallejo, California; and one could include Stamford's Harbor Point in that category – all had failed efforts before success. It takes an extraordinary location, deep pockets and sometimes more than one developer for such large projects to proceed. On the other end of the scale, very small projects such as gas stations (discussed in Chapter 6) may sometimes simply seem not worth the time and costs, but they can be successful and perhaps lead to more significant opportunities. With the real estate bubble behind us, it appears that projects in the middle – say in the broad range between $20 million and $200 million, like those in Trenton, Portland and elsewhere seem to now have a greater likelihood of success. Large projects are often accomplished in phases, reducing capital requirements and building towards an area-wide revitalization.

Meeting the market: all development projects must satisfy market demand and attract end users. Waterfront redevelopments must utilize the water as an asset rather than an access barrier. The mix of uses, attractive design and events that all bring a community back to its waterfront have all been key to successful projects. There are projects, often large or unique that effectively make their own market, but far more make the most of market trends, mining them for the most advantageous features ranging from residential unit size to amenities to hot retailers.

Positive government: as cited many times, redevelopments involve multiple government agencies at all levels, certainly as regulators and often as champions and financiers. Most successful projects, especially complex examples such as waterfront brownfields and affordable housing, also have government as a partner. While there are relatively few direct government grants to for-profit developers, there are often government loans, environmental assessment grants, loan guarantees, financial mechanisms, land cost write-down, and infrastructure support. As noted throughout, public-private partnerships are increasingly the way urban redevelopment gets done. While formerly more common in large cities and large projects, this trend has spread throughout North America. The public sector brings not only approval action but sometimes tax advantages, infrastructure and some financing to these projects, while the developer provides capital, expertise, construction capabilities and whatever else needs to be done. How the community is engaged, with both the government and private development aspects of development, is critical. Public entrepreneurs, such as economic developers, brownfield specialists, non-profit and community leaders also lead as well as support many public-private efforts. Cities that had functional, timely approval processes were more likely to be successful. Certainly leadership committed to redeveloping waterfronts and brownfields matters, including having the right personnel in the planning and economic development departments was an important factor.

Feasibility and financial strength – multiple financing streams: making use of not just conventional equity and debt, but also government grants and loans, specialized project funds such as those for community amenities are all utilized. Managing such financing, having sufficient funds at the necessary moments, from private and various levels of government sources, is clearly a significant challenge. The majority of successful projects include a market component, parts of the project that can be conventionally financed based upon future rents or sales. These projects may also have a financing gap as discussed in Chapter 8, which may be filled by government funds from federal, state or local levels in various forms: grants, loans, real estate tax abatements, tax credits or use of TIF. Some projects, often those that have significant park or cultural features, may receive significant charitable or foundation support. Managing these multiple funding streams, sequencing what can be done as funding is available, is a key to a successful urban redevelopment.

Design quality: redevelopment requires a highly sophisticated, multitalented design team. The design needs to work for the community. This may well involve historic preservation or more generally a sense of authenticity and the funky feel of a neighborhood. The design team also needs to be fully cognizant of environmental constraints, public access, amenities, viewsheds and ecological opportunities, community context, as well as all the normal market and cost considerations. The developer needs to select and support the unique design, a shared vision that is financially viable and will maximize return on investment.

See the big picture and the individual impact. An urban redevelopment is not just a hotel or a shopping center; it is part of an urban community in the way it is designed, the amenities that are offered and the environmental and social impact. It is place-making, in a city, not just a building. Providing new homes, lively and safer streets, a cleaner environment improves residents' lives, preserving community and ameliorating gentrification remains a conundrum. Urban redevelopment often brings more good things than bad, but the human impact of displacement has to always be a foremost consideration.

The redevelopment of the former Rheingold Brewery, in what had been the largely industrial and quite tough Bushwick section of Brooklyn, New York, is a good example of the redevelopment process, including design. Started in the 1850s, this brewery survived prohibition and, with clever marketing such as the famous Miss Rheingold competition, flourished, until overwhelmed by market changes and national competitors. The plant closed in 1978, was demolished and the site lay fallow and contaminated for the next two decades. The German-American Bilateral Working Group, sponsored by US EPA and their German counterpart, visited the Rheingold site and came up with a concept plan. There was strong local and political support, and by the end of 2003, a local community organization (Ridgewood-Bushwick Senior Citizens Council), with strong political support, had become the project champion. A partnership with a private development partner, the Bluestone Organization, was formed and the first phase of the redevelopment planned. Bluestone Rheingold Partnership Homes had 140 units and a traditional townhouse and apartment format,[7] and much of the financing came from New York City and state

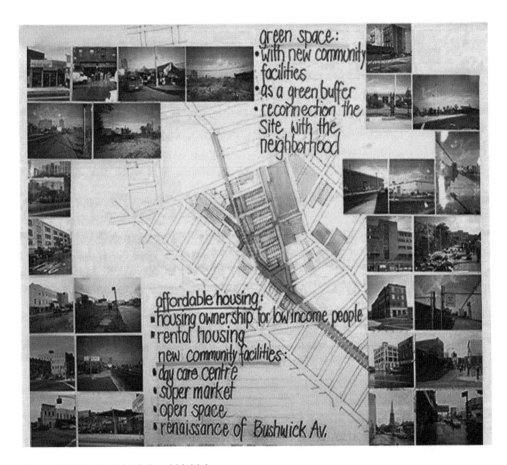

Figure 10.2 Bushwick Rheingold initial concept
Source: US EPA German-American Bilateral Planning Group Design Charette.

affordable housing and brownfield programs. By 2015, there had been several development phases, some with different private partners, and over 800 units had been built. The latest phase features an imaginative and cutting-edge sustainable design by ODA with a garden roof. This history reflects how urban redevelopment succeeds: a site opportunity, a vision, community engagement, political support, developers, availability of financing, environmentally responsible design – over more than a decade.

Pulling all the threads together, having the right people and making the right decisions in each specialty leads to successful redevelopment. Leadership counts, and successful projects have not only a champion, but also many supporters. Most often there is a key individual who, with many allies, takes on a project and makes it happen. This means mastering each issue, often with the help of specialists, and persevering as many urban developments take years and in some cases decades and must overcome many obstacles.

The project must be financially feasible, if not, it will be a failed effort. Whether privately funded or a public-private project, whether there are large institutions involved or not, the project has to be funded, construction paid for and have an ongoing income sufficient to operate and pay off mortgage or other debt. Creating the financial model and finding both

Figure 10.3 Bushwick Rheingold Brewery latest phase
Source: ODA.

equity and debt, as well as whatever government funding is needed to fill a gap, is critical. Many urban redevelopments are privately funded, but many others involve some form of government support, from historic tax credits, brownfield incentives to tax increment financing.

Communication and community engagement: a successful urban redeveloper must be in constant touch with many of the stakeholders; government officials, community leaders, advocates of different types, as well as investors, lenders, regulators and prospective tenants. In the new world of social media, this includes websites, blogs and many other internet tools, in addition to the traditional media and the still necessary face-to-face meetings. The Detroit Riverfront Conservancy, heir to numerous design award-winning projects, continues to hold events, send out e-blasts and promote what is an active, downtown waterfront community.

Urban redevelopment should take full advantage of unique intown locations. A large proportion of projects are at the core of the city; historic preservation and funky authenticity can be great assets. Waterfronts are often prime redevelopment opportunities, but they require storm resilience, ecological responsibility, public access requirements and specialized expertise. A key to urban redevelopment is seeing how such advantageous but costly locations can be maximized by creative reuse.

Finding the right team, the specialized expertise of the urban designer, attorneys, architect, engineers, environmental experts and financial advisors, is also critical. It is the whole team and the proper leadership who help bring a project to fruition. Urban redevelopments need to contribute to what Charles Landry describes as the "distinction, variety and flow" of a great city.[8]

Urban redevelopment is complicated and often takes longer than all hope. Big projects have to be capable of surviving changes in political regimes, market cycles, setbacks, lawsuits and more. A project must have the capital and other resources and the time to succeed. Starting a substantial redevelopment is an act of courage.

There needs to be active community engagement for urban redevelopment to succeed. Top down urban renewal, despite often good intentions, often did not work. These developments are in neighborhoods whose residents observe, comment, vote and speak. Social and environmental equity are vital in today's urban world. The local community must receive appropriate benefits from the project, such as good jobs, affordable housing and public amenities.

The good news is that the wind is at the back of urban redevelopment; there are strong markets, available financing, some government and community support and expertise to deal with even the thorniest issues. As demonstrated in many locations throughout North America, successful urban development projects work at many levels, functionally serving the market, enhancing communities, improving sustainability, waterfront access and resilience that feature fully protective remediation and, most importantly, reflect local aspirations. To accomplish successful projects requires developers to think strategically and to use techniques that reflect the unique nature of each of these projects, maximizing financial, aesthetic and community benefits. The best thinkers about cities; economists, lawyers, scientists, architects and planners have found that indeed, every city is different and urban redevelopment is complex. That does not mean that important analytic work is not being done, or that we cannot learn from one another. It does mean that many the factors, from historic preservation to environmental remediation to affordable housing and more need to be considered. So this book is for the practitioners, those who struggle to find the right tools, use the appropriate disciplines, and make the most of the unique, complex opportunities in urban redevelopment. What this report hopefully does is provide insights so that more developers and their teams are well prepared to deal with the wide range of complicated issues in waterfront brownfields. There remain many urban development opportunities with enormous potential, awaiting the developer and the community with the right skills, strengths, perseverance and a little bit of luck. Done right, urban redevelopment has a bright future.

Notes

1 Miles, Mike E., Laurence M. Netherton and Adrienne Schmitz, *Real Estate Development – 5th Edition: Principles and Process*, Urban Land Institute, 2015.
2 Peiser, Richard and David Hamilton, *Professional Real Estate Development: The ULI Guide to the Business, 3rd Edition*, 2012.
3 Lewin, Roger, *Complexity: Life on the Edge of Chaos*, University of Chicago, 1999.
4 Kahneman, Daniel, *Thinking, Fast and Slow*, Farrar, Straus and Giroux, 2011.
5 Lewis, Michael, *The Undoing Project*, W.W. Norton, 2016.
6 Watkins, Stefan D., "The Impact of Brownfield Reclamation on Surrounding Land Values and Crime," Department of Resource Analysis, Saint Mary's University of Minnesota, Minneapolis.
7 Hersh, Barry, *Renewal and Redevelopment*, 2014.
8 Landry, Charles, *The Creative City: A Toolkit for Urban Innovators*, Earthscan, 2008.

Bibliography

Abelson, Edward and William M. Seuch, "Redeveloping Contaminated Real Estate: Realizing Brownfield Opportunities," Brownfields Lawyer, 1999.

Adler, Jerry, *High Rise: How 1,000 Men and Women Worked Around the Clock for Five Years and Lost $200 Million Building a Skyscraper*, HarperCollins, 1993.

Alexander, Christopher, Sara Ishikawa, Shlomo Angel and Murray Silverstein, *A Pattern Language: Town, Buildings, Construction*, Oxford University Press, 1977.

Alexander, Christopher, *A City is not a Tree*, Architectural Record, 1965.

Angel, Shlomo, *Planet of Cities*, Lincoln Institute of Land Policy, 2012.

Breen, Ann and Dick Rigby, *Waterfronts: Cities Reclaim Their Edge*, Macmillan, 1994.

Breen, Ann and Dick Rigby, *The New Waterfront: A Worldwide Urban Success Story*, McGraw-Hill, 1996.

Breen, Ann and Dick Rigby, *Intown Living: A Different American Dream*, Praeger, 2004.

Caro, Robert, *The Power Broker: Robert Moses and the Fall of New York*, Random House, 1974.

Castells, Manuel, *The Informational City: Economic Restructuring and Urban Development*, Blackwell, 1992.

Chakrabarti, Vishaan, *A Country of Cities: A Manifesto for Urban America*, Metropolis, 2013.

Clark, Gregory, *The Business of Cities*, Routledge, 2017.

Conn, Steven and Max Page, *Building the Nation: Americans Write About Their Architecture, Their Cities and Their Landscape*, University of Pennsylvania Press, 2003.

Cunningham, Storm, *The Restoration Economy*, Berrett-Koehler Publishers, Inc., 2002.

Czerzniak, Julie, *Formerly Urban: Projecting Rustbelt Futures*, Princeton Architectural Press, 2013.

Dunham-Jones, Ellen and June Williamson, *Retrofitting Suburbia: Urban Design Solutions for Redesigning Suburbs*, John Wiley & Sons, 2011.

Eichholtz, P., Nils Kok and J.M. Quigley, "Doing Well by Doing Good? Green Office Buildings," *The American Economic Review*, 100 (5), 2492–2509.

Ewing, Reid and Otto Clemente, *Measuring Urban Design: Metrics for Livable Places*, Island Press, 2013.

Florida, Richard, *The Rise of the Creative Class*, Basic Books, 2002.

Florida, Richard, *The Rise of the Creative Class Revisited*, Basic Books, 2012.

Florida, Richard, *The New Urban Crisis*, Basic Books, 2017.

Gallagher, Leigh, *The End of the Suburbs*, Penguin, 2013.

Garreau, Joel, *Edge City: Life on the New Frontier*, Anchor Books, 1992.

Glaeser, Edward, *Triumph of the City*, Penguin, 2010.

Graham, Wade, *Dream Cities: Seven Urban Ideas That Shape the World*, HarperCollins, 2016.

Harnick, Peter and Ryan Donahue, "Turning Brownfields into Parks," *Planning*, December 2011.

Heller, Gregory L. and Ed Bacon, *Planning, Politics, and the Building of Modern Philadelphia*, University of Pennsylvania Press, 2013.

Hersh, Barry, "Defensible Space: A Personal Reflection," *CityScape*, HUD, Volume 1 Number 3, 1996.

Hersh, Barry, "Real Estate Tax Policies and Brownfield Redevelopment," Lincoln Institute of Land Policy, Cambridge, MA, 2002.
Hersh, Barry, "Brownfields by the Bunch," *Brownfields Renewal*, February 2010.
Howland, Marie, "What Makes for a Successful Brownfield Redevelopment, Three Baltimore Case Studies," University of Maryland, 2002.
Hudnut, William, Tom Murphy, Ed McMahon, Michael Beyard, John McIlwain, Robert Dunphy and Steve Blank, *Changing Metropolitan America*, Urban Land Institute, 2008.
Iannone, Donald T., *Redeveloping Urban Brownfields*, Land Lines, 1995.
Inam, Assam, *Designing Urban Transformation*, Routledge, 2014.
Jacobs, Jane, *Death and Life of Great American Cities*, Vintage Books, 1961.
Kahneman, Daniel, *Thinking, Fast and Slow*, Farrar, Straus and Giroux, 2011.
Kahr, Joshua and Michael C. Thomsett, *Real Estate Valuation and Market Analysis*, John Wiley & Sons, 2005.
Kelly, Hugh, *The 24 Hour City: Real Performance Not Just Promises*, Routledge, 2016.
Kotkin, Joel, *The New Suburbanism*, The Planning Center, 2006.
Kotkin, Joel, "The Geography of Aging: Why Millennials are Headed to the Suburbs", *New Geography*, 2013.
Kramer, Anita, Terry Lassar, Sara Hammerschmidt and Mark Federman, *Building for Wellness: The Business Case*, Urban Land Institute, 2014.
Landry, Charles, *The Creative City: A Toolkit for Urban Innovators*, Earthscan, 2008.
Leary, Michael E. and John McCarthy, *The Routledge Companion to Urban Regeneration*, Routledge, 2014.
Lewin, Roger, *Complexity: Life on the Edge of Chaos*, University of Chicago Press, 1999.
Lewis, Michael, *The Undoing Project*, W.W. Norton, 2016.
Lynch, Kevin, *The Image of the City*, MIT Press, 1960.
Miles, Mike E., Laurence M. Netherton and Adrienne Schmitz, *Real Estate Development Principles and Process, 5th Edition*, Urban Land Institute, 2015.
Meyer, Peter, *State Initiatives to Promote Redevelopment of Brownfields and Depressed Urban Areas: An Assessment of Key Features*, United States Department of Housing and Urban Development Office of Policy Development and Research, 1999.
New Partners for Community Revitalization, "Evaluation of the New York State Brownfields Opportunity Area Program," NYU Wagner Capstone (Tyler Gumpwright, Rose Martinez, Rachel Cohen, Sam Levy, Javier Garciadiego, Prof. Michael Keane), May 2016.
Newman, Oscar, *Defensible Space*, Macmillan, 1972.
Norquist, John O. *The Wealth of Cities, Revitalizing the Centers of American Life*, Perseus, 1998.
Owen, David, *Green Metropolis*, Penguin, 2009.
Peirce, Neil R. and Curtis W. Johnson with Farley M. Peters, *Century of the City: No Time to Lose*, Rockefeller Foundation, 2008.
Peiser, Richard and David Hamilton, *Professional Real Estate Development: The ULI Guide to the Business, 3rd Edition*, Urban Land Institute, 2012.
Platt, Rutherford H., *The Humane Metropolis People and Nature in the 21st-Century City*. Amherst: University of Massachusetts in Association with Lincoln Institute of Land Policy, Cambridge, 2006.
Porter, Michael, "On Competition," *Harvard Business Review*, 1980–1995.
Rappaport, Nina, *Vertical Urban Factory*, Actar, 2016.
Romer, Paul, with Luis Rivera-Batiz, "Economic Integration and Endogenous Growth," *Quarterly Journal of Economics CVI*, May 1991.
Rose, Jonathan F.P., *The Well-Tempered City*, HarperCollins 2016.
Rybczynski, Witold, *Last Harvest: How a Cornfield Became New Daleville*, Scribner, 2007.
Sagalyn, Lynne, *Times Square Roulette, Remaking the City Icon*, MIT Press, 2001.
Sarni, William, *Greening Brownfields: Remediation Through Sustainable Development 1st Edition*, McGraw-Hill, 2010.

Shepard, Mark and Michael Stubbs, *Urban Planning and Real Estate Development/2nd Edition*, Taylor & Francis, 2012.
Smart Growth America, *(Re)Building Downtown: A Guidebook for Revitalization*, Washington, DC, 20036; 2015.
Smith, Andrew, *Events and Urban Regeneration: The Strategic Use of Events to Revitalise Cities*, Routledge, 2012.
Sobel, Lee, Steven Bodzin and Ellen Greenberg, *Greyfields into Goldfields: Dead Malls Become Living Neighborhoods*, Congress for the New Urbanism, June 2002.
Sutton, Stacey, A., *Urban Revitalization in the United States: Policies and Practices*, Columbia University, 2008.
Sykes, Peter and Hugh Sykes, *Urban Regeneration: A Handbook*, SAGE, 2008.
Taleb, Nassim, *The Black Swan*, Random House, 2007.
Taleb, Nassim, *Antifragility*, Random House, 2012.
Townsend, Anthony, *Smart Cities: Big Data, Civic Hackers, and the Quest for a New Utopia*, W.W. Norton, 2013.
Watson, Donald, *Design for Flooding*, John Wiley & Sons, 2011.
West, June A. and Rebecca Goldberg, "Repurposing American History: Steel Production Ends in Bethlehem, Pennsylvania," *Harvard Business Review*, 2011.
Wheeler, Stephen M. and Timothy Beatley, *Sustainable Urban Development Reader/2nd Edition*, Routledge, 2008.
Whyte, William H., *The Social Life of Small Urban Spaces*, Project for Public Spaces, 1980.
Wolfe, Tom, *A Man in Full*, Farrar, Straus and Giroux, 1998.

Index

Page numbers in *italics* denote tables, those in **bold** denote figures.

24/7 gateway cities 13

active design 82
adaptive reuse 23–30
Advance Real Estate 130
aerial photography 39
AeroFarms, Newark, New Jersey 119
affordable housing xvi, 56, 114, 116–18, 124, 128, 149, 154; Denver Transit Oriented Development (TOD) 75, 76; Detroit 145; financing 116; and inclusionary zoning 116–17, 120; Low-Income Housing Tax Credits (LIHTC) 101, 116, 123, 128; and master planning 120; mixed-use 117–18; New York City 117–18, 155; public-private partnerships 116, 117
agriculture, urban 119, 126, 143–4
AIG Global Real Estate 137
air quality 95
Alexander, Christopher 47
Altman, Andy 106
Amazon headquarters, Seattle 64
Ambler Boiler House, Pennsylvania 23
Americans with Disabilities Act (1990) 38, 50
apartments 10, 126; live/work 118
Arcadia REIT 125
Arlington County, Virginia, Rosslyn Ballston Corridor 62, 63
Arringon, G.B. 7, 56
artists' housing 118
arts and culture 54, 82, 84–5
Asheville, North Carolina 87
Atlanta: Atlantic Station 137–8, 154; Peachtree Center 17; Piedmont urban park **81**
Atlantic Station, Atlanta 137–8, 154
automobiles 44, 47

Bacon, Edmund 5, 33
Baltimore 2–5, 154; affordable housing 118; downtown projects 3, 4; East Baltimore Development Initiative (EBDI) 4; Inner Harbor redevelopment 2–3, **4**, 25; Johns Hopkins University 4, 85, 128; Oriole Park, Camden Yards 3; Port Covington neighborhood 5; Sandtown-Winchester neighborhood 4
Barrington-Leigh, Christopher 8
Battery Park City, New York City 31, 38, 141
Bay Area Rapid Transit District (BART), San Francisco 64, 65–6
Beasley, Larry 58
Bedrock Real Estate Services LLC 146
Bell, Michael 98, 109
Berger (builders) 5
Bethesda, Maryland 130
Bethlehem Redevelopment Authority, 29, 30
Bethlehem Steel 28–30, 146; Hoover-Mason Trestle 29; SteelStacks **29**, 30
bicycles and bikeways 44, 59, 82
Biederman, Dan 133
big data 39, 40, 70
bio-remediation 104, 109
block grid hierarchy 44
Bloomberg, Michael 141
blotting process, Detroit 147
Bluestone Rheingold Partnership Homes 155
Blumenfeld, Hans 52
Bonaventure Hotel, Los Angeles 17
Booker, Cory 124
Boston Redevelopment Agency (BRA) 31
boundaries 42, 48–9
Breen, Ann and Dick Rigby, 179
Bricktown, Oklahoma City 17
Bromberg, David 84
Brookfield Properties 138
Brookings Institution 131
Brooklyn Academy of Music (BAM) 84–5
Brooklyn Heights, New York City 28
Brown, Jerry 115
Browner, Carole 89
Brownfield Development Areas program, New Jersey 102

Brownfield Economic Development Initiative (BEDI) 100
Brownfield Opportunity Areas Program, New York State 95, 101–2
brownfields 13–14, 28, 89–112; area-wide/corridor approach 94–5, 101–2; Atlantic Station, Atlanta 137–8; Camden, NJ 123–4; community perspective on remediation of 93; Detroit 146, 147; Dry Gulch Stream restoration, Lakewood, Colorado 86; environmental liability protections and insurance 94, 104, 137; Environmental Protection Agency (EPA) program xv, xvi–xvii, xviii, 13, 89; gas station redevelopment 110–11; Greenway, Ranson, West Virginia 86; institutional controls 104, 152; land values 91–2; New York City 99, 101, 141; presumptive remedies 104; public/private partnerships 93; remediation *see* remediation of brownfields; self-certification 104; site environmental assessment 94, 100; tax abatements for remediation 94, 117, 134, 147; technology and remediation of 104; *see also* waterfront brownfields
Brundtland Commission 33
Bruner Cott & Associates 84
Bryant Park, New York City 82, 133
Bucholtz, Marjorie 89
building codes 42
Building Information Modeling (BIM) 39–40, 153
Building Information Systems 150
Bureau of Economic Research 89
Burgess, Ernest W. 47
Burnett, James 86
Burnham, Daniel 81
buses 44
Business Improvement Districts (BIDs) 133

Cabrini Green, Chicago 115
Cahn, Amy Laura 5, 6
Callahan, John 28, 29
Camden, New Jersey 102, 121, **122**, 123–4
Canadian Mortgage and Housing Corporation 114
capital markets *see* real estate and capital markets
Carnegie Mellon University, Collaborative Innovation Center 17
Carter, Majora 114
casino developments: Sands Casino, Bethlehem, PA 28, 30; Vancouver 59
CCLR (California Center for Land Recycling) 18
Cecil Hotel, Harlem, New York City 117–18
Centerpoint Properties Trust 125
Central Business Districts (CBDs) 10, 11, **12**

champion(s) 150, 153, 154
Charleston, South Carolina 22
charrette approach 153
Cherokee Investment 94, 102, 123
Chicago 49; Cabrini Green 115; Oaklawn neighborhood 56
Chicago Lake Front 81
Chicago, University of 85
Chicago: Growth of a Metropolis 49
Christman Company 146
Church of Jesus Christ of Latterday Saints (LDS Church) 128
civic pride/accomplishment 34
Civil War battlefields 21
Clark University, Worcester 85
Claridge House, Detroit 147
class stratification xvi
Clean Water Act (1972) 94
Cleveland 119, 134; Health and Technology Corridor 131; Slavic Village Development 126
climate factors 40, 41, 53; *see also* storm events; wind
coastal zone management programs 101
Cocoziello, Peter 130
color 52
Commercial Mortgage-Backed Securities (CMBS) 126
commercial real estate markets 126; property prices 12, **13**; vacancy rates 10
communication 153, 156–7
communities: and brownfield development 93; engagement of 36–7, 120–1, 136, 149, 154, 156–7; as leaders in development 2
Community Development Block Grant program 2
Community Development Finance Agencies 131
community gardens 119
community spirit 54
commuting 10
Complexity: Life on the Edge of Chaos (Lewin) 151
complexity of redevelopment flow chart 151, **152**, 152
computer aided design 39
Conklin + Rossant 86
construction 45
construction technology 40
Cook County, Illinois 117
Cook, Rodney Mims 138
Corktown, Detroit 147
costs 42, 43; life cycle 53
creative economy **9**
crime prevention 55–7
crowdfunding 121, 125, 126, 136
cultural facilities 54, 82, 84–5
cultural factors 36–7
Cuomo, Andrew 115

Dallas: Deep Ellum 84; Klyde Warren Park **83**
Daniels, Ronald 4
Darden, Tom 102
de la Uz, Michelle 121
Dearborn, Michigan 36
Deep Ellum, Dallas 84
Defensible Space (Newman) 50–1, 55–6
demographic change 8
Denver Dry Goods building **27**, 27
Denver Regional Transit District (RTD) system 71–2, 73, 76, 77, 78
Denver Transit Oriented Development (TOD) 71–2, 73–9; affordable housing 75, 76; Arvada Stations 74, 77–8; Blueprint Denver (2002) 74; City Center Englewood (2000) 74; City of Denver TOD initiative (2006) 74; City of Denver TOD Manager (2015) 76; Denver Livability Partnership (2011) 76; Denver TOD Fund (2015) 75; Denver TOD Strategic Plan (2006) 74; FasTracks plan 71, 72, 73; Housing Development Assistance Fund (2011) 76; light rail system 71–2, 73, 76–8; RTD TOD Pilot Projects (2010) 76; RTD TOD Strategic Plan (2010) 76; TOD Strategic Plan Update (2014) 76; TRD TOD Assessment (2015) 76; Urban Land Conservancy (ULC) 75
Denver Union Station (DUS) 71, 72, 77, 78
Department of Housing and Urban Development *see* HUD
Department of Transportation (DOT) 150
design charrettes 153
design quality 155–6
Detroit 14–15, 119, 126, 142–7; affordable housing 145; Blight Removal Taskforce 147; blotting process 147; brownfield projects 146, 147; Claridge House 147; Corktown 147; Dodge Fountain 143, 144; downtown development 145–6, 147; Eastern Market 147; Far Eastside Plan 144; property tax reform 147; Red Wings arena 145; Renaissance Center (RenCen) 14, 15, 142–3; streetcar system 145; streetlighting 145; Theater District 143; Tiger Stadium 143; urban agriculture 119, 143–4; waterfront redevelopment 144–5
Detroit Lighting Authority 145
Detroit Riverfront Conservancy 144–5, 157
Detroit Venture Partners (DVP) 146
Diller, Barry 140
disabled access 33–4, 50
Discovery Green, Houston 88
displacement of residents 26, 75, 84, 113, 115, 149, 155
DNAPL (dense non-aqueous phase liquids) 90, 103
Dodge Fountain, Detroit 143, 144

DOT (Department of Transportation) 150
downtowns 5; Baltimore 3, 4; Detroit 145–6, 147
Doxiadis, Constantine 5
Dranoff Properties 121
driving 10
drone surveillance 41
drone technology, and urban design 39
Dry Gulch Stream restoration, Lakewood, Colorado 86
Duany, Andres 38, 137
Duggan, Mike 145, 147
DUMBO (Down Under the Manhattan Bridge Overpass), Brooklyn, NY 84
Durham Performing Arts Center, Durham, North Carolina 86–7, **87**

earthquakes 42, 70
easements 42
East Baltimore Development Initiative (EBDI) 4
Eastwick Friends and Neighbors Coalition 6
Eastwick, Philadelphia 5–6
EB-5 visa financing 126, 136
economic clusters 131
economic development xv–xvi, xvii, 2, 130–3
Economic Innovation Group 13
edge cities 130
edges 48–9
EDR Insight 14
educational institutions 85, 128
Eisenhower, Dwight 5
electricity 70
Embarcadero Center, San Francisco 17
emergency centers 41
eminent domain 2, 100, 101, 115, 123, 128
employment trends 8, **9**
energy 53–4; conservation 96; management systems 40; micro-grids 70; renewable 41, 70
Energy Star 53
Enterprise Communities 3
entry/exit points 40, 50
Enviro-Stewards 24
environmental codes 42
Environmental Data Resources 89
Environmental Impact Statements 38, 136
environmental insurance 94, 104, 129, 137
environmental issues 33, 38, 82, 151–2; *see also* brownfields
environmental justice xiii, xviii, 5, 99, 101
Environmental Protection Agency (EPA) 91, 93, 104, 150; Brownfield Assessment Grants 89, 100, 101; brownfields program xv, xvi–xvii, xviii, 13, 89; Superfund resuse program 129; XL program 137
equity financing 125, 126, 149
Evergreen Brickworks, Toronto **23**, 24
Exploratorium, San Francisco 15

Falls Church, Virginia 130
family size 8
Faneuil Hall, Boston 25
Fannie Mae (Federal National Mortgage Association) 114, 126
feasibility 155, 156
Federal Emergency Management Agency (FEMA) 41, 69
Federal Home Loan banks 114
Federal Housing Agency 114
ferries 105–6
FHLB cooperative bank 116
Fields, Timothy 28, 89
financial crisis (2007—2010) 2
financing 37, 43, 125–6, 128–9, 131–3, 133–4, 149, 150, 152, 155, 156; affordable housing 116; megaprojects 136; waterfront brownfields redevelopment 100–2, 108, 117; *see also* tax credits; Tax Increment Financing (TIF)
fire codes 42
flooding/flood control 41, 69, 106
floorscapes 52
Florida, Richard 8, 13
Flushing, Queens County, New York 36
focus, creation of 50
Ford, Rob 34, 114
foreclosures 126, 134
form *see* urban design form
Fort Trumbull, New London, Connecticut 101
Foster, Norman 47
Freddie Mac 126
Freeman, John 97, 107
freeways 16
Friedman, Adam 26
Fuller, Buckminster 41
Fulton, Bill 5
functional relationships 48
Furstenberg, Diane von 140

Gallagher, Leigh 6
Garczynski, Lynda 89
gardens, community 119
Garfield Traub Development 86
Garreau, Joel 130
gas station redevelopment 110–11
Gas Works Park, Seattle 85
Gehry, Frank 47
General Electric 12
gentrification 26, 75, 84, 93, 113, 114, 115, 120, 140–1
Geographic Information Systems (GIS) 39, 40, 99
George, Henry 119
Georgetown, Washington, DC 25
Ghirardelli Square, San Francisco 16
Gilbert, Dan 145, 146
Glaeser, Edward, *Triumph of the City* 28, 47, 142
globalism, and urban design 31

Goldman Sachs 127, 147
Goldschmidt, Neil 17–18
government: and economic development xv–xvi, xvii–xviii; positive 154
government funding 128–9, 136, 149, 155; *see also* tax abatements; tax credits; Tax Increment Financing
Government Sponsored Entities 126
Grace, Christopher 98
Grand Central Partnership 133
Grand Central Station, New York City 22
Granville Island, Vancouver 16, 58
Gray, Freddie 4
Great Recession 2, 126
Green Corridors 86
green roofs 60, 70
green standards 70, 82; *see also* LEED standards
Greenwich Village, New York City 25, 28
grey-water systems 70
Grosvenor Americas 59
Gruen, Victor 43

Hadid, Zaha 47, 140
Hantz Woodlands, Detroit 143–4
Harbor Point Stamford, CT 97, 98, 99, 102, 105, 106–8, 107, 154
Harrison, New Jersey 130
Haussmann, Georges-Eugène, Baron 1
healthcare services 117–18
heat island effect 41, 82
heating, ventilation and air conditioning (HVAC) systems 40, 70, 95
Heckendorn Shiles 23
Heckscher, August 46
hedge funds 136
High Line, New York City 138, **139**, 140, 141–2, **142**
High Tech Districts 131
historic preservation 21–30, 150, 157
Historic Sites and Monuments Board of Canada 21
historic tax credits 21, 22–3, 27, 101, 128, 149
Historic Valdese Foundation 126
Hoboken, New Jersey 10, 130
Home Depot 131
Home Loan Bank 116
Hoover-Mason Trestle, Bethlehem Steel 29
HOPE VI program 115, 123
Horton Plaza, San Diego 15
housing 113–18, 131; apartments 10, 118, 126; HOPE VI program 115, 123; multifamily **11**; property prices 10; public 2, 56, 114–15, 123; senior citizen 114, 115, 116, 123; single-family 7, 8, 10, **11**; SRO (single-room occupancy) 118; subsidies 101, 114, 116; transitional/supportive 117–18; workforce 117; *see also* affordable housing

Houston 47; Discovery Green 88
Hoyt, Homer 47
HUD 115, 116, 150; Brownfield Economic Development Initiative (BEDI) 100; HOPE VI program 115, 123; Section 8 voucher program 2, 114, 115, 116
Hudson Yards, New York City 31, **32**, 43, 55, 138, **139**, 140, **142**, **144**, 154
human scale 51–2
Huntington, Henry 62
Hunt's Point neighborhood, New York City 114
Hyatt Regency San Francisco 17
hydroponic farming 119

Illinois, University of 128
Image of the City (Lynch) 47
inclusionary zoning 116–17, 120
Independence Mall, Philadelphia 33
Independence National Historic Park, Philadelphia 50
individually-focused programs 11–12
industrial buildings, renovation of 25–7, 28–30
Industrial Development Agencies 131, 132, 134
Industrial Development Bonds 116, 136
Industrial Revenue Bonds (IRB) 131, 134, 149
infrastructure xvi, xviii, 7, 23, 31, 34, 36, 42, 43–5, 47, 69–70, 117, 127, 128, 129, 130, 131, 136, 150
Innovation Districts 131
institutions, role in redevelopment 85, 128
insurance companies 125, 136
interface development 49
Investment in Affordable Housing (IAH) 114

Jacobs Center for Innovation, San Diego 121
Jacobs, Jane 2, 28, 48, 56
Jacoby Development 137
job creation 130, 150
Johns Hopkins University 4, 85, 128
Johnson-Marshall, Percy 46
joint ventures, waterfront brownfields redevelopment 102 102
juxtaposition 51

Kahn, Louis I. 54–5
Kahneman, Daniel, *Thinking, Fast and Slow* 151
Kelo v. *New London* (2005) 2, 101
Kildee, Dan 147
Klyde Warren Park, Dallas **83**
Kohn Pedersen Fox 55
Koolhaas, Rem 119
Korman Companies 5, 6
Kotkin, Joel 7
KPF architects 138
Kraemer Design Group PLC 146
Krens, Thomas 84
Kresge Foundation 147

Lackawanna, New York 28
Lakewood, Colorado, Dry Gulch Stream restoration 86
land 126–7
Land Trusts/Banks 118–19, 134, 143, 153
land use planning/regulation 7–8, 37–8
Landers, Brad 121
Landry, Charles 157
Le Corbusier 114, 119
leadership 153–4; and waterfront brownfield redevelopment 97–8
LEED standards 33, 53, 70, 82, 93, 96, 137–8
Leinberger, Chris 7
L'Enfant, Pierre Charles 33
Lewin, Roger, *Complexity: Life on the Edge of Chaos* 151
Lewis, Michael, *Undoing Project, The* 151
Liberty Property Trust 121, 123
Lichtenstein, Harvey 84
life cycle considerations 53–4
light: artificial/night-time 51, 53, 56, 70; natural 51
light rail systems 16, 44; Denver 71–2, 73, 76–8; Portland, Oregon 17–18, 62–3, 64, 68–9; Rosslyn Ballston Corridor, Arlington County 62–3; Seattle streetcar project 63, 64
Lincoln Land Institute, Cambridge, Mass. 119, 147
living alone 8
local materials, use of 33, 45
loft developments 24, 25–6
Los Angeles 36, 47, 62; Bonaventure Hotel 17
Low-Income Housing Tax Credits (LIHTC) 101, 116, 123, 128
Lynch, Kevin, *Image of the City* 47

McGalliard Falls Park, Valdese, NC 126
McGee, Dean A. 86
Malloy, Dannell 98, 107
malls: Horton Plaza, San Diego 15; Independence Mall, Philadelphia 33
man-made catastrophes 41
Mandell, Rick 10
market demand 154
Markowitz, Marty 85
Massachusetts Museum of Contemporary Art (MASS MoCA) 84
master palnning *see* urban design plan/master plan
medical institutions 85, 128
megaprojects 136–48, 153; Atlantic Station, Atlanta 137–8, 154; *see also* Detroit; Harbor Point Stamford; Hudson Yards, New York City
metrics 45–6
Metrotech, Brooklyn, NYC 84–5
Meyer, Peter 92

Miami-Dade County 38
Milan, Milton 123
Miles, Mike E. 149
Millard-Ball, Adam 8
Millspaugh, Martin 3
Milwaukee 18
Mirontschuk, Victor 106
Moinan Group 138
Montreal 33, 114
Morristown, New Jersey 130
mortgage-backed securities 126
Moses, Robert 2, 34
Moynihan, Daniel Patrick 113
multifamily housing **11**
Murase Associates 86
Muscogee, Oklahoma 17
Myriad Botanical Garden, Oklahoma City 86

National Association of Home Builders 7, 10
National Environmental Policy Act (1970) 98
National Historic Landmarks list 21
National Historic Preservation Act (NHPA) 21
National Museum of Industrial History 29
National Register of Historic Places 21
National Trust for Historic Preservation 21
Native American settlements 22
natural disasters 40, 41
natural gas 70
natural light 51
nature 53
New Eastwick Corporation 5
New Market Tax Credits 101, 116, 119
New Rochelle, Westchester County, New York 7
New York City: affordable housing 117–18, 155; Battery Park City 31, 38, 141; Brooklyn Academy of Music (BAM) 84–5; Brooklyn Heights 28; brownfields projects 99, 101, 141; Bryant Park 82, 133; Business Improvement Districts (BIDs) 133; Department of Housing and Preservation 120; DUMBO (Down Under the Manhattan Bridge Overpass) 84; Economic Development Corporation 120; Fifth Avenue, Brooklyn 121; Grand Central Station 22; Greenwich Village 25, 28; High Line 138, **139**, 140, 141–2, **142**; Hudson Yards 31, **32**, 43, 55, 138, **139**, 140, **142**, **144**, 154; Hunt's Point neighborhood 114; loft developments 25, 26; Metrotech Center, Brooklyn 84–5; Planning Commission 140; Rheingold Brewery, Brooklyn 155; Seward Park, Lower East Side 120; SoHo neighborhood 25–6; South Street 25; Times Square 45, 49; Willet's Point project 101, 137; Yonkers 55–6
New York Municipal Art Society 22
New York University, Center for Urban Science and Progress (CUSP) 141

Newark, NJ: AeroFarms 119; Teachers Village 118
Newman, Oscar 115, 129; *Defensible Space* 50–1, 55–6
NIMBY (Not In My Backyard) reaction 36
noise 95
Norquist 18
Nouvel, Jean 140

Oaklawn, Chicago 56
office space 10; flexibility and adaptability 45
Oike Place Market, Seattle 16
Oklahoma City: Bricktown 17; Myriad Botanical Garden 86
Old City of Quebec 22
Olmstead, Frederick Law 81
Onassis, Jacqueline Kennedy 22
O'Neill, J. Brian 97
Ontario Science Centre 15
open spaces 50, 81–2; *see also* parks
Opportunity Funds 126
over-the-road tractor trailers 44
Oxford Properties 138

Paris 1
Park, Robert E. 47
parking 44, 70, 72
parks 37, 81–2, 85–7, 87–8; *see also individual parks entries*
Parsons, Anne 145
pattern 52–3
Pavia, Michael 108
paving materials 49
paving patterns 53
Payment in Lieu of Taxes (PILOT) 132, 134
PCBs (polychlorinated biphenyls) 90, 103
Pearl District, Portland 68–9
Peachtree Center, Atlanta 17
Pedersen, William 55
pedestrian access 43, 50
Peebles, Brad 109
Pei, I.M. 86
Peiser, Richard 149
Penn's Landing, Philadelphia 17
Pennsauken, New Jersey 102, 123, 124
Pennsylvania, University of 85, 128
pension funds 125, 136
perspective 49
Philadelphia: Eastwick 5–6; Independence Mall 33; Independence National Historic Park 50; Penn's Landing 17; Spruce Street Harbor Park 87; urban farming programs 119
Philadelphia Land Trust 118
Philadelphia Redevelopment Agency 6
photography, aerial 39
physical analysis 36
phyto-remediation 104

Piano, Renzo 47
Piedmont urban park, Atlanta **81**
Pioneer Square-Skid Row Historic District, Seattle 22
Pittsburgh 131
place-making 33, 48, 56, 81, 84, 95, 155
Plank, Kevin 5
Plaza of Nations, Vancouver 59
Plensa, Jaume 87
pocket neighborhoods 57–8
Port Covington, Baltimore 5
Porter, Michael 130–1
Portland, Oregon 25, 72, 73, 98; Central City Plan (1988) 69; Downtown Plan (1972) 69; light rail systems 17–18, 62–3, 64, 68–9; Pearl District 68–9; zoning 69
Portman, John 17
Post Properties, Atlanta 130
Pratt Institute, Brooklyn, NYC 85
preservation, historic 21–30
project scale 154
property prices: commercial 12, **13**; housing 10
Property Reserve, Inc. 128
Pruitt Igoe, St Louis 115
public health 82
public housing 2, 56, 114–15, 123
public performance, opportunities for 54
public-private partnerships xvii, 2, 31, 128, 149, 154; affordable housing 116, 117; brownfield redevelopment 93; megaprojects 136; waterfront brownfields redevelopment 102–3

Quebec 114; Old City 22

racial stratification xvi
Radburn, New Jersey (NJ) 5
Rails to Trails program 150
Ranson, West Virginia, Greenway 86
Rappaport, Nina 26
RCRA *see* Resource Conservation and Recovery Act
Real Capital Analytics and Walkscore 12
real estate and capital markets 10, 125–35; *see also* commercial real estate markets
real estate crowdfunding 121, 125, 126
real estate data systems 40
Real Estate Investment Trusts (REITs) 125
real estate marketing 129
Related Companies 138
religious practices 36
remediation of brownfields 19, 28, 85, 89–90, 91, 92, 93, 94–5, 96, 99, 100, 101, 103–5, 111, 116, 129, 152; bio-remediation 104, 109; phyto-remediation 104; tax abatements for 94, 117, 134, 147; triad approach 104, 150
renewable energy 41, 70, 150
Renaissance Center, Detroit 14, 15, 142–3

Renn, Aaron 17
rental values 10
Requests for Proposal (RFPs) 38, 101, 102, 103, 117, 120, 149
resilience 31, 33, 40–2, 70, 95
Resource Conservation and Recovery Act (RCRA) 28, 89, 90
Reston, Virginia 31
retail projects 15–16, 131; New Market Tax Credit 116; Vancouver 59; *see also* malls
revitalization xv–xvi, xvii
Revolutionary War battlefields 21
Reynolds, david 5
RFPs *see* Requests for Proposal (RFPs)
Rheingold Brewery, Brooklyn, New York City 155
Richmond, California 7
Riley, Joe 22
River Action, Inc. 120–1
Rock Ventures LLC 146
Rockefeller Center 25
Rocket Street, Little Rock, Arkansas 57–8
Roman Catholic Church 35
roof gardens 119
roofs, green 60, 70
Rose, Jonathan F.P. 27, 127
Rosslyn Ballston Corridor, Arlington, Virginia 62, 63
Rouse, Jim 3
RTC (Resolution Trust Company) 126
Rudofsky, Bernard, *Streets for People* 47

Safdie, Moshe 51
safety 34, 40–2, 50–1, 56–7, 129
St Louis, Pruitt Igoe 115
Salt Lake City, City Creek Center 128
Sampson, Robert J. 12
San Diego: Horton Plaza 15; Jacobs Center for Innovation 121
San Francisco 131; Bay Area Rapid Transit District (BART) 64, 65–6; Embarcadero Center 17; Exploratorium 15; Ghirardelli Square 16; Hyatt Regency 17
Sands Casino, Bethlehem, PA 28, 30
Sandtown-Winchester, Baltimore 4
Saperstein, Harriet 144
satellite technology 39
Savannah, Georgia 22
scale 48; human 51–2; project 154
Schaefer, Donald 3
schools 129
Schwartz, Robin 146
science parks 17
sculpture 54
Seattle 33; ferries 105; Gas Works Park 85; Oike Place Market 16; South Lake Union (SLU) neighborhood 64, 84; streetcar project 63, 64

Section 106 Review 21
Section 8 voucher program 2, 114, 115, 116
security 33, 34, 40–2, 50–1, 56–7, 129
self-gentrification 114
senior citizen housing 114, 115, 116, 123
Serbinski, Ted 146
service economy 8, **9**
Seward Park, New York City 120
sewer water 69, 70
shade and shadow 51, 82
Show Me A Hero (TV miniseries) 55
sidewalks 49
sight lines 49
signage 53
Silicon valley 131
Simon Fraser University 58
single-family housing 7, 8, 10, **11**
site orientation 48
skyscrapers 55
Slavic Village Development, Cleveland 126
Small Business Liability Relief and Brownfields Revitalization Act (2002) 100
Smart Cities 70
smart growth 2, 69, 95
Smithsonian Institute 29
SOBRO (South Bronx Overall Redevelopment Organization) 18
social equity 113–24
social justice xv–xvi
SoHo, New York City 25–6
South Lake Union (SLU) neighbourhood, Seattle 64, 84
spatial flexibility and adaptability 45
sprawl 8, 16, 95
Spruce Street Harbor Park, Philadelphia 87
Stamford, Connecticut, Harbor Point 97, 98, 99, 102, 105, 106–8, **107**, 154
Stanislaus, Mathy 121
Stanley Park, Vancouver 60
Stapleton, Colorado 31
State Historic Preservation Offices/Officers 21, 27
Stein, Clarence 5
Stein, Robin 107
steps 52
Stevens, Rod 14–18
stigmatized locations 129
storm events 42, 70
storm protection 41, 42, 82
storm water 69, 70, 82, 93, 96
stratification of populations xvi
street furniture 52
street hierarchy 44
street walls 52; openings 52
streetcar systems 16, 63–4, 68–9, 145
streets 49
Streets for People (Rudofsky) 47

subsidies, housing 101, 114, 116
suburban development 1, 7, 8, 10, 11
subways 44
Sullivan, Louis 55
Summit Realty Advisors 23
sun/sunlight 51, 53
"superblocks" 2, 44
Superfund (CERCLA) regulations 89, 90
superimposition 51
surveillance cameras 56
sustainability 31, 33, 45, 53–4, 69–70, 95
Sustainability Community Initiative 100, 101
synergy 154; remediation/redevelopment 103–5
Szostak Design 86

target hardening 56
Taubman Centers, Inc. 128
tax abatements 117, 132, 133–4, 155; brownfield 94, 117, 134, 147
tax credits 155; historic 21, 22–3, 27, 101, 128, 149; New Market 101, 116, 119; *see also* Low-Income Housing Tax Credits (LIHTC)
Tax Increment Financing (TIF) 5, 28, 37, 108, 116, 117, 128, 132–3, 134, 136, 149
tax liens 134
tax revenues 130
taxi and livery vehicles 44
Teachers Village, Newark, NJ 118
team-building 97–8, 157
technology 131, 146; and brownfield remediation 104; and urban design 39–40
Tenancy in Common (TIC) 125
terrorist attacks 57, 70
texture 52–3
Thinking, Fast and Slow (Kahneman) 151
This Old House (TV series) 27
Thompson, James and Jane 3
Thor Equities 125
TIF *see* Tax Increment Financing
Tiger Stadium, Detroit 143
Times Square, New York City 45, 49
Tishman Speyer 138
Toledo, Ohio 154; brownfield redevelopment 97–8, 102–3, 108–10
Toledo-Lucas County Port Authority 108–9
Toll Brothers 130
Toronto 114; Business Improvement Districts (BIDs) 133
Toronto adaptive reuse examples: Evergreen Brickworks **23**, 24; North Toronto Station 24–5, **25**; Toy Factory Lofts 24
Trammell Crow 78
Transit Adjacent Development (TAD) 72
Transit Oriented Development (TOD) 72–3, 95; equitable 75; *see also* Denver Transit Oriented Development
transition 51

transportation 10, 16, 43–4, 45, 62–80, 129, 131, 150; Bay Area Rapid Transit District (BART), San Francisco 64, 65–6; bicycles and bikeways 44, 59, 82; buses 44; ferries 105–6; freeways 16; light rail systems *see* light rail systems; streetcar systems; over-the-road tractor trailers 44; private automobiles 44, 47; subways 44; taxi and livery vehicles 44; Tysons Corner Virginia 64, 66–7; vans and pickup trucks 44; and waterfront brownfields redevelopment 99; *see also* Denver Transit Oriented Development (TOD)
Trenton, New Jersey 94, 98, 99, 103, 104, 154
triad remediation 104, 150
Trinity College, Hartford 85
Triumph of the City (Glaeser) 28, 47, 142
Tversky, Amos 151
Tysons Corner Virginia 64, 66–7

Undoing Project, The (Lewis) 151
United Nations (UN) 33, 35
urban agriculture 119, 126, 143–4
urban design 31–61; civic aspects of 34; community factors 34, 36, 54; financial factors 37; globalism and 31; guidelines 33, 38–9; local materials and 33, 45; parameters 36–9; physical analysis 36; process 35–6; resilience and 31, 33, 40–2; safety/security considerations 33, 34, 40–2, 50–1, 56–7; sustainability and 31, 33, 45, 53–4; team members 35–6; technology and tools for 39–40; and zoning and land use regulation 37–8
urban design elements 51–7; color, texture and pattern 52–3; "fixtures of the plaza" 52; graphic signage and way-finding 53; human scale 51–2; life cycle, energy and sustainability considerations 53–4; lighting 53; nature 53; public community spirit 54; public performance, art and sculpture 54; street furniture 52; street walls and floorscape 52
urban design form 46–51; ascent/descent 49–50; boundaries and edges 48–9; entry/exit and pedestrian access 50; focus, creation of 50; functional relationships 48; hierarchy 50; juxtaposition/superimposition/transition 51; landmark, history and fabric 48; natural light, shade and shadow 51; perspective and sight lines 49; project composition 47–8; scale 48; site orientation 48
urban design plan/master plan 34, 35, 38, 42–6; and affordable housing 120; and construction 45; goals and objectives 42–3; metrics 45–6; program 43; review process 43; spatial flexibility and adaptability 45; urban fabric and infrastructure 43–5
Urban Land Institute 149

Urban Renewal 2, 114–15, 128
urban sector theory 47
Urban Vertical Factory **26**, 26–7
US Fish and Wildlife 69
utility networks 45

Valdese, North Carolina 126
value engineering 42, 53
Vancouver 33, 58–60; bikeways 59; casino project 59; Coal Harbor 58; False Creek 58–9; Granville Island 16, 58; green roof project 60; Plaza of Nations 59; rental housing 59; retail projects 59; The Rise project 59; Simon Fraser University 58; Stanley Park 60; viaducts removal 59; Woodward's project 58
Vancouver Convention Center West 60
vans and pickup trucks 44
variegated network model 47
vehicle miles per capita **10**
vehicular access 43; control 41
vertical connection issues 49–50
vertical farming 119
vision 154
visualization 39–40

walkability 12, 72, 129
Wallace, Davis 3
walls, street 52
Washington, DC 33, 72, 73; Georgetown 25
water: sewer 69, 70; storm 69, 70, 82, 93, 96
water quality 94–5
water resources 41, 41–2
waterfront brownfields 91, 92, 93, 96–110, 130, 151, 152–3, 154; approval strategies 98–100; Camden, NJ 123–4; environmental assessment 100; financing 100–2, 108, 117; Harbor Point Stamford, CT 97, 98, 99, 102, 105, 106–8, **107**; joint ventures 102; leadership and team building 97–8; mixed-use development 105–6; public/private partnerships 102–3; remediation/redevelopment, synergy between 103–5; site acquisition 102–3; stakeholders 99–100; Toledo, Ohio 97–8, 102–3, 108–10
waterfront redevelopment 69–70, 151, 154, 157; Baltimore 2–3, **4**, 25; Detroit 144–5; *see also* waterfront brownfields
Westbank Projects Corporation 58
wetlands protection 105
White House Council on Environmental Quality initiative 99
Whyte, William H. 56, 82, 113, 133
Willet's Point, New York City 101, 137
Wilmington, Delaware 84
wind 42, 53
Wine, Kathy 121
Wolfe, Jordan 147

workforce housing 117
workplace trends 8, **9**, 10
Wright, Henry 5

Yale University 85
Yonkers, New York City 55–6

zoning 7–8, 37–8, 52, 127, 149; form-based 38; inclusionary 116–17, 120; Manhattan West Side redevelopment 140; Portland, Oregon 69; transects 38
Zucker, Paul 51–2
Zyber-Platek, Elizabeth 38, 137

Taylor & Francis eBooks

Helping you to choose the right eBooks for your Library

Add Routledge titles to your library's digital collection today. Taylor and Francis ebooks contains over 50,000 titles in the Humanities, Social Sciences, Behavioural Sciences, Built Environment and Law.

Choose from a range of subject packages or create your own!

Benefits for you
- Free MARC records
- COUNTER-compliant usage statistics
- Flexible purchase and pricing options
- All titles DRM-free.

Benefits for your user
- Off-site, anytime access via Athens or referring URL
- Print or copy pages or chapters
- Full content search
- Bookmark, highlight and annotate text
- Access to thousands of pages of quality research at the click of a button.

Free Trials Available
We offer free trials to qualifying academic, corporate and government customers.

eCollections – Choose from over 30 subject eCollections, including:

Archaeology	Language Learning
Architecture	Law
Asian Studies	Literature
Business & Management	Media & Communication
Classical Studies	Middle East Studies
Construction	Music
Creative & Media Arts	Philosophy
Criminology & Criminal Justice	Planning
Economics	Politics
Education	Psychology & Mental Health
Energy	Religion
Engineering	Security
English Language & Linguistics	Social Work
Environment & Sustainability	Sociology
Geography	Sport
Health Studies	Theatre & Performance
History	Tourism, Hospitality & Events

For more information, pricing enquiries or to order a free trial, please contact your local sales team: **www.tandfebooks.com/page/sales**

 | The home of Routledge books | **www.tandfebooks.com**